The Exit Side of the Universe

The Real Beginning

Rolando Rivero

Designed by Linda F. Hawkins

ISBN: 978-0-9912047-0-0

"The suppression of uncomfortable ideas may be common in religion and politics, but it is not the path to knowledge, and it has no place in the endeavour of science."

Carl Sagan

Table of Contents

Introduction

My name is Rolando Rivero. I am a visionary and believe I have uncovered the most important discovery in the history of mankind. My discovery is about the beginning of our universe — the beginning of our world and the beginning of us.

According to scientific experts, the Big Bang created everything in our universe. The theory that it was created by nothing at all seems very scientific. Because even with all the great scientific minds and the mathematical data and the hard work involved in researching the beginning of our universe, this theory still does not answer the HOW and WHY we were created. The Exit Side of our Universe is my story. In it I will show you pictures, proof and the ordinary common sense theory that our universe was created by other energy. And, that these energies came from a multiverse.

Atheists might claim that this theory proves the methodical geometry of the multiverse and its energy sources are easily explained without the presence of faith in a deity.

For thousands of years, mankind has struggled to find answers to the great mystery of when and how our universe began. Religious teachings would have us believe that there was nothing – no air, no gravity, no mass, no water – nothing but darkness, until God decreed there should be earth and light. The Book of Genesis in the Christian Bible goes into great detail to explain just how God created the Earth and all its inhabitants. What is curious to me is that ancient civilizations like the Norsemen, Native Americans, Hindus, the Chinese and others all have a similar creation story. All state that in the beginning there was nothing. All claim that a supreme being, all with the unique yet similar story, created the Earth.

All around the world, ancient man built extraordinary monuments. And although separated by vast distances and thousands of years, researchers believe that a secret code, embedded in these megalithic monuments, may reveal the truth behind the connection between numerous ancient civilizations.

Is it a coincidence that many of the greatest minds in human history have received information from an otherworldly source? Or could it be possible that Socrates, Leonardo da Vinci, Nikola Tesla, Srinivasa Ramanujan, Albert Einstein and other geniuses used altered states of consciousness to access a realm that contains the entire knowledge of the universe?

Is this knowledge of the universe some sort of code for creation? Could there be a creation code? All the sciences, chemistry, physics, astronomy, geology and biology, all contain formulas – codes – that create the elements and actions of our world. Could the creation code be hidden within sacred writings of the Bible and other religious stories?

Theories abound regarding the creation of life on Earth, from primordial soup, clay being formed into life, silicon evolved into plant life. Perhaps it is all true. Consider that silica is in every human connective tissue: our bones, our teeth, our glands and organs. Small organisms like sponges depend on silica for their structural strength. Clay is mostly silicon. Earth contains clay. God shaped man from the Earth.

Hydrogen is the number one particle in our universe while silicon is the eighth most abundant. Could the high levels of hydrogen on early Earth have created the perfect environment to allow silicon crystals to develop into life forms? We know these early life forms evolved into grasses and flowering plants, but could they have been the beginnings of other life forms as well? Fossilized bee-like wasps with pollen still intact on their hind legs believed to date from the Cretaceous Period, 100 million years ago, helped to create the variety of plants we have today.

Deoxyribose sugar found in honey contains the four nitrogen based sugars: Guanine, Cytosine, Adenine and Thymine which are necessary to create the proteins to sustain life. Bees have an intelligence encoded into their genes that is passed on to every bee. One so intelligent that every honeycomb cell is a perfect hexagon shape. No one can recreate it by hand. Is it God, who inspires in bees the knowledge and ability to do this? Bees are vital to our survival. Over 100,000 species go extinct each year. Since life began on Earth, five major mass extinctions have led to large and sudden drops in the biodiversity of species.

Regardless of where we came from, where we may go from here, you can be assured that there is a plan for tomorrow. Our very human nature is geared for survival and for those who place their faith in a higher power. The Christian Bible states that "For I know the plans I have for you, declares the Lord, plans for welfare and not for evil, to give you a future and a hope." Jeremiah 29:11.

Before I continue my story, I'm going to tell you a little bit about myself. From the time I was a little kid I would have ideas and thoughts about our outer space, our universe. And these thoughts were sometimes quick. And in my thoughts I would have visions of motions of light: stars, planets, galaxies, black holes, universe, multiverse. All would travel around in my thoughts in motion very quickly, experiencing these visions in a first person state of mind, going through the motions of traveling in space. As I got older, I was beginning to understand that this was a gift. The information on space and our universe on TV, radio, and books would seem wrong. But I was not sure, because I am not a scientific expert nor have I ever studied the contents of outer space or the universe.

But time would pass, experts in their scientific fields like astronomers, astrophysicists would stand corrected to new findings, as well as confirm my vision as correct. I think that almost everyone has different kinds of special gifts. Now how would one explain knowing outer space, our universe, multiverse? How bizarre is that? No skills, no experience, never studied, no scientific data. These thoughts would come and go, and the older I got, the more I would understand. When I was 30 years old, the visions would surface again. But this time something was different. This time my visions and feelings were stronger. Now there were audible voices. My body would have chills and goosebumps all over. It was an incredible great euphoric feeling. Almost like there was something I needed to explain, but not ready. I realized that all these correct thoughts had to be coming from God. I have attended church on and off since I was about four years old. I am not what you call a constant religious person. But make no mistake, I do believe in one God, Jesus Christ, and the Holy Spirit. And this one God is the creator of the multiverse, our universe, our solar system, Earth, Humans, and more. He's the creator of everything.

I also believe in being good, righteous, courteous, great family foundations and human exceptionalism.

Even though I don't go to church every week, I walk with God every minute of the day. So I know that this special gift comes from God. As years passed, I would remember these thoughts of a multiverse, ingrained in my brain. At 46 years old on April 23, 2012 at 6 PM, I was sitting on my front porch, when this voice in my head with these visions and thoughts of the multiverse, became extremely strong. This time it was so much different. I walked in to my house so dazed that I could hardly see my surroundings, not realizing I had actually entered my home. A voice thought said, "You need to create this now, you need to show the world a new secret discovery, it is time."

So I answered God's call to create this; however, I wondered how to start and where to begin. God said to me, "Don't worry my son I will guide you." So without wasting a moment I quickly jumped into my truck and drove to the nearest hobby shop. I walked all the way to the back of the store and picked up about 36 Styrofoam circular shapes. I then continued to search the store finding colored toothpicks, which I quickly purchased and then returned home. At the back of my house is a 12 x 20 shed that I call my man cave. Once inside, I quickly went to work and in about 10 minutes I had put together 33 Styrofoam circles using these colored toothpicks to hold it together. I created some sort of weird shape. I stared at this shape trying to figure out what it was. Clearly, one circular shaped Styrofoam piece in the center with 12 other Styrofoam circular shapes covering the center one, with 20 other Styrofoam circular shapes covering 20 points from the center Styrofoam circular shape. At that moment, I knew exactly what I had created and needed to do. I had created a model 33 multiverse with our universe being in the center. Now it was time to dissect the workings of all these universes. I backtracked by pulling off one Styrofoam circular shape at a time. The outer Styrofoam ball layer was held together by 20 green toothpicks. The inner 12 Styrofoam shapes were held together by yellow toothpicks. All these toothpicks were embedded into the center Styrofoam circular shape. And in the center all these toothpicks were touching. This is how our Big Bang began. It was created by 32 multiverse feeding energy to one point. 13.7 billions of years ago the Holy Grail, the God particle was in motion and the Big Bang had begun. But what put these multiverse in motion? In the real beginning, there was a big, really big, massive dark sphere. In this one massive dark sphere with a whole lot of pressure coming from the center, is an outer edge. And on this outer edge almost simultaneously God's pulsating energy was feeding this dark pressure sphere.

God's pulsating energy would push into this dark massive sphere and because this dark sphere had a lot of pressure it would push back on God's pulsating energy and create a crash or clash and heat rhythm constantly pushing and pulling. This was on the outer edges of this massive dark sphere whose substance was similar to silicone or gases like silicone gas. Craters were beginning to appear. Almost simultaneously a triangle, or pyramid, or tetrahedral-like shape started emerging. And this crashing and constant movement of pushing and pulling defined these pyramid-like shapes. Soon because the tetrahedrons and craters were in the circular rotations of the outer 20 spheres the multiverse began. Because of these 20 multiverse particles and God's pulsating energy, the inner 12 multiverse spheres were created and the cycle continued with these 32 multiverse particles and God's pulsating energy creating our number 33 in the center of the universe. And of course, with all 32 multiverses having their own big bang. Cold or hot, up or down, in or out, cause or effect, push or pull, crash or rotating – all these constant changes – the perfection is what permits these multiverses to continue. It all made sense to me now. I had to create a book about my theory of the 33 multiverses establishing the beginning of the basic map, by mapping north, south, east, west and 33 multiverses in motion from all its points. Also inbetween these multiverses are what I call a neutral space area. So I needed to show the work, by making molds with colors. I went to a couple of different stores and purchased silicone, clay, cast, Styrofoam, plaster, spackling, glue, spray paint colored blue, yellow, green, orange. Within 2 months time I had completed 20 molds of the inner workings of a three dimensional 33 multiverse.

I took 332 pictures in details from every angle as well as a video during the creation of this. And there are people who know I created this. Note God and the Holy Spirit are the reason why I am able to have these visions to create – again I am no expert in the field. Although this phase is completed I am not done yet. There is a reason for my book, the beginning of map-making of our 33 multiverses.

My map and my book are only the beginning. My job has just begun, my job is to show the world my new discovery, my new secret of the multiverse. I am to show the world so the world's scientific experts can help dissect and define and build on my new secret discovery. I need help from all kinds of people in the scientific world. I need help because it's going to take experts lots of years to perfect the multiverse. Future generations will use this map of the multiverse for travel. And those generations will need to be very well prepared.

The reason that God chose me is because I am not an expert. You see, in this world, there is a very big separation between religion and science. Religion and science are constantly arm wrestling. To me, they're both the same. But let's face it, before there was science, there was religion. From these religious thoughts science was created. If God chose a scientific expert to discover the secrets of the multiverse, some people would deny it and say, "He knows that because of his scientific expertise." But coming from me, a non-scientific person, it is no doubt believable.

YHVH

YHVH

God's Creation

I saw no temple in it, for the Lord God the Almighty and the Lamb are its temple. And the city has no need of the sun or of the moon to shine on it, for the glory of God has illumined it, and its lamp is the Lamb. *Revelations 21:22-23*

There is the holiest of places called Heaven. Here is where God lives. Heaven is massive, endless. The Bible says that Heaven is surrounded by the light of the lamb. This light is a completely different kind of light. This light is different than anything we have on earth or in our universe. So in the beginning God opened in the center of heaven an area or space. He stretched and stretched and kept stretching almost to be endless. To us it would seem infinite. And God stretched the center of Heaven. The lamb of light was compressed to the extreme, like a sponge, trying to compress back.

In the beginning God created the heavens and the earth. And the earth was formless and void, and darkness was over the surface of the deep; and the Spirit of God was moving over the surface of the waters. Then God said, "Let there be light", and there was light. *Genesis 1:1-2*

The Bible tells us that God created the Heavens and the Earth – that there was nothing until he created it. No form, no light. Nothing until he decreed it. But how did he do it? What action did he take to create it? Mankind has searched for that answer for thousands of years. And even today, with our advanced scientific knowledge, we are only beginning to understand the entire story of creation. For all our telescopes and microscopes, our observatories and laboratories, our research and studies, we still are at a loss for a concrete answer. All we have are theories. Scientific theories that often collide with religious views. Religious views that often disagree with scientific facts. And yet there are points within each where there is agreement.

Seven books of the Bible mention God creating and stretching the heavens. The creation story takes place in just one week, yet these seven books span a time period of 500 years, beginning in 1000BC with Psalms to approximately 518BC in the book of Zechariah. Science wants us to believe there was a big instantaneous bang and suddenly the universe was born and that was it. But something tells me there is more to the story. This book will share my theory on the rest of the story, which I feel God has called me to share with you.

God created a massive, really massive, dark void – the center of his heaven. Since there is no written document, there is no name for it, but I'm going to call this – Rolando Rivero's dark pressurized void. And everywhere in this dark void which seems empty, it is full of compressed dark matter. Although you can't see it, dark matter is there, confined in its sphere edges. This void was created so light can travel, because darkness was created for light. We think that light was created for darkness. You see we think there isn't anything faster than the speed of light – wrong – darkness is faster than light. For example, any kind of light has a beginning source of power that triggers light to travel outwards to a certain length. But when you cut the source of power, that length of travel speed to become dark is faster. So God said let there be light and this light was good.

Before the Big Bang, our universe was a dark sphere. But it was far from empty. You see, our universe has an edge holding in this pressurized dark void. And almost simultaneously, 32 pulsating string energies penetrated our universe edges with a perfect mathematical pattern and gravity was born. Together with creation of dark energy, these 32 energies and gravity and dark energy were in a tug of war for a very, very long time. Because these 32 intelligently mathematical fanned patterned pulsating string energy overwhelmed dark void, dark energy by using gravity to their advantage until it they reached our universe's center point and this massive big bang was created. Albert Einstein's E=mc2 on steroids! I call this Rolando's E=mc32.

The full creation of yellow and green multi-verse will be explained in later chapters.

Ancient societies in ages past believed in the mythological creations of gods and deities. Some humans were considered to have special powers which enabled them to tap into new ideas and theories through meditation, dreams, thoughts and visions. Even in modern times there are those whose "out of the box" thinking has brought new insights and inventions to our world. Some of these people were mathematicians, scientists, physicists, philosophers, historians and artists. Some were just common everyday people like you and me. So it is possible that I am able to harness this power and bring new knowledge just as the icons of history have done. Perhaps, it could even be possible for you to do the same.

The Coexistence of Science and Religion

This centric multi-verse theory does not make any attempt to dispute the long established religious concepts in today's society. The theory is open-ended and leaves room for personal speculation as to the energy sources beyond the system and what lies on the exit side of the multi-verse. If one so chooses, a deity could be inserted as an explanation to the unknown energy sources that are shown to have effect on the multi-verse. Religious institutions may argue that these factors are evidence in support of an omniscient outer entity. On the contrary, atheists might claim that this theory proves the methodical geometry of the multi-verse and its energy sources are easily explained without the presence of faith in a deity. This being said, new information can either unite a society or divide it. Science and religion have the capability of working together, if the communities so choose, as to not hinder advancements in technology and further developments of civilization.

For thousands of years mankind has struggled to find answers to the great mystery of when and how our universe began. Religious teachings would have us believe that there was nothing – no air, no gravity, no mass, no water – nothing but darkness, until God decreed there should be earth and light. The Book of Genesis in the Christian Bible goes into great detail to explain just how God created the Earth and all its inhabitants. What is curious to me is that ancient civilizations like the Norsemen, Native Americans, Hindus, the Chinese and others all have a similar creation story. All state that in the beginning there was nothing. All claim that a supreme being, all with their unique yet similar story, created the Earth. The Chinese creation myth of Pan Gu and Nü Wa is curiously similar to the Big Bang theory. According to this story, Pan Gu was a giant who grew inside an egg-shaped cloud containing all the matter of the universe. Inside this egg all the matter of the universe was swirling in chaos. For 18,000 years Pan Gu slept in the egg and grew so large that one day when he awoke and stretched he broke the egg and released all the matter of the universe.[1]

According to Dr. Fritz Schaefer, Graham Perdue Professor of Chemistry and Director of the Center for Computational Quantum Chemistry at the University of Georgia, there are five traditional arguments for the existence of God. Schaefer states that:

1. The cosmological argument: the effect of the universe's existence must have a suitable cause.
2. The teleological argument: the design of the universe implies a purpose or direction behind it.
3. The rational argument: the operation of the universe, according to order and natural law, implies a mind behind it.
4 The ontological argument: man's ideas of God (his God-consciousness) implies a God who imprinted such a consciousness.
5. The moral argument: man's built-in sense of right and wrong can be accounted for only by an innate awareness of a code of law – an awareness implanted by a higher being.

[1] Jan Walls and Yvonne Walls (translators and editors), 1984, Classical Chinese Myths: Hong Kong, Joint Publishing Company, 135 p. (BL1825.C48 1984)

Until the 20th century, the conventional wisdom of science was that the universe was static and eternal. That it did not have a specific beginning or cause. Even Albert Einstein had philosophically resisted the idea that there was a beginning. Over time, he began to accept a necessity for a beginning, although he never accepted the idea of a personal God. Such resistance stems from the first argument for the existence of God. In other words, everything that has a beginning must also have a cause. So if the universe began to exist, then what was its cause?

However, since the 1920's, that theory has been challenged with new knowledge that creates more questions about the beginnings of our universe. Russian-born scientist, George Gamow suggested that the Big Bang was an intense concentration of pure energy. And also that all the galaxies should be rushing away from one another as a result of the explosion.

By 1965, Arno Penzias and Robert Wilson, both from the Bell Telephone laboratories had convinced many scientists that the Big Bang theory was valid through their observations of microwave background radiation. COBE (the NASA Satellite Cosmic Background Explorer) was launched November 18, 1989 and carried three instruments: a Far Infrared Absolute Spectrophotometer (FIRAS) to compare the spectrum of the cosmic microwave background radiation with a precise blackbody, a Differential Microwave Radiometer (DMR) to map the cosmic radiation precisely, and a Diffuse Infrared Background Experiment (DIRBE) to search for the cosmic infrared background radiation.[2] COBE radically changed our understanding of how the universe began. Its precise mapping and measurement of the oldest light in the universe – the cosmic microwave background, provided confirmation of the Big Bang theory of the origin of the universe.

This confirmation of the Big Bang also makes the idea that God created the Universe more plausible than at any time in the last 100 years. Still not all scientists, including Fred Hoyle and Geoffery Burbidge, agree with the Big Bang theory and believe in the now obsolete steady-state hypothesis that new matter is continuously created as the universe expands. Hoyle's theory was that new matter formed in what he called a "creation field", which had negative pressure in order to be consistent with the conservation of energy and drive the expansion of the universe. Burbidge suggested that the steady-state theory supports Hinduism rather than Christianity because it supports their belief in the endless cycles of creation, preservation and dissolution of the universe. [3]

Creationist and founder of science-faith think tank Reasons to Believe, Hugh Ross, uses scientific evidence to argue for the truth of Christianity. This comparison was found on his blog, "Scientists say the universe had a beginning, is stretching, and will eventually come to an end. The Bible says the universe had a beginning, is stretching outward, and will eventually come to an end."

Surely the answers lie somewhere between Religion and Science. Again we can look at creation stories of the ancients. Religious writings and viewpoints from of old going as far back as the artwork on the cave walls depicting a larger than life being, some theorists going so far to describe them as alien beings. Philosophers and early men of science all

[2] http://science.nasa.gov/missions/cobe/
[3] http://carm.org/hinduism-beliefs

searching for understanding. All wanting an answer to the beginning of the universe. So where to begin? I, Rolando Rivero, believe, through my visions and messages from God, that I can shed light on the mystery and at the very least open the minds of others who will take my visions on to the next level of discovery.

Alien secret code. Stonehenge in England, peyote, the great pyramids of Egypt, tales of Teotihuacan, Mexico and Machu Picchu, Peru. All around the world, ancient man built extraordinary monuments. And although separated by vast distances and thousands of years, researchers believe that a secret code, embedded in these megalithic monuments, may reveal the truth behind the connection between numerous ancient civilizations. When you look at where the ancients built their monuments, it is clear that they were marking the places where the power of the earth could be harnessed. So the question really is can we still maintain that all of these sites were simply chosen by accident? Or is there far more to it than that?

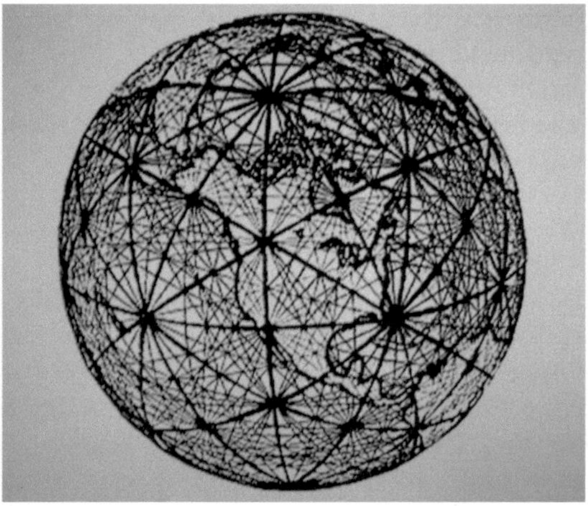

The evidence clearly suggests that these cultures were doing something on a global scale. Interacting, creating a pattern which is really unique and lost to us right now. Some researchers have discovered a strong connection between sacred structures and powerfully charged areas of the globe, where electromagnetic energy gathers. Today, electromagnetic energy waves provide all visible light, radio and cellular communications, as well as microwaves and x-rays. Energy scientifically is the property that technically allows things to work. So we think of things like magnetic fields, electric fields, gravity, things moving, heating things up, those are all forms of energy. And the one thing that we do know is that the earth has a magnetic field and that magnetic field connects different parts of the earth. And that strange anomalies occur, like gravitational anomalies and magnetic anomalies. And even like a knowledge to do with this duration of visions. And if you sort out all the marks on the map you will see that they seem to form geometric grids, known as the world grid. Ancient astronaut theorists believe extraterrestrials constructed their monuments on energy lines that, when mapped and connected, created a significant pattern. The concept of the world grid is that the earth is like a giant crystal and energy is flowing around it at little nodules, and energy paths are intersecting and moving all over the world.

By the year 650 B.C., a formidable technology had arisen in the Mediterranean, Greece in particular. People had learned how to survey land and navigate by the stars: they had a sophisticated metallurgy. And what is important to my story is that they had a detailed knowledge of the positions of stars and planets for making calendars and assorted predictions.

Socrates was the founding father of Western philosophy. His contemporaries described the philosopher as being strange in many ways, including spending hours each day consumed by his own thoughts. One of the most stunning examples of Socrates' day-dreaming actually happens while on campaign as a soldier at the siege of Potidaea (432-429). He displayed an extraordinary endurance that his peers may have found odd and suspicious and puzzling. We also think that a genius is an eccentric but I think some of their eccentricities come from the fact that they really are in touch with other realities. In the case of Socrates, he said he was able to access a demon. Doubtful, for Socrates is not some evil creature. But ancient Greece had this subcategory of gods who really were talking to human beings explaining the mysteries of life, explaining why we were here. Is Socrates so-called demon evidence that the founder of Western philosophy was mentally unstable? Or could it be the proof that he was in contact with an extra terrestrial intelligence? Or God was sending him messages from a distance realm in the cosmos? It is too easy just to say that yes, Socrates was nuts, or that he was sick, schizophrenic or hearing voices. Was that demon some kind of messenger? Some kind of bridge between him and the realm of the metaphysical? Was Socrates' practice of deep meditation a connection to God? Could there be other geniuses with similar claims from a higher realm while in a trance like state?

Plato (c.430-c.354 B.C.), the fourth century BC Greek philosopher and mathematician, taught that the basic structure of the Earth evolved from geometric shapes known as Platonic solids. Plato left a tantalizing clue for the possibility of the first mention of the world grid. He basically described the earth as being created from 12 pentagonal faces, and 20 points or vertices on the surface. When you join these together and mark them on a map, there is a kind of geometric formation that appears between them. Plato wrote that there was a world soul which he described as a sphere that was composed of 120 equal identical triangles.

Plato discovered 2500 years ago that man was able to capture and use the Earth's energy at power surges around the world to help build their megalithic monuments. There are fascinating ideas that they were actually using this for anti-gravitational purposes. They were actually able to lift antiquary and move the huge megalithic blocks great distances around the planet. So we have the question of what this something is. What it has to do with this grid actually harnessing energy. And how were they able to use this energy to move and construct the sites.

In the 6th century B.C. Democritus of Abdera suggested that atoms were the key to a simple universe, thus began the search for the truth about the creation of Earth and the workings of our universe. Yet, the concept of atoms was not embraced until the 19th century.

In the Northwestern jungle of Cambodia stands the largest Hindu temple in the world. Angkor Watt, built in the 12th century, its design is meant to represent Mount Meru, the mythical mountain that exists in the cosmos. The Hindu believed it was the center of the metaphysical universe. Numerous stone carvings adorn the walls of the temple depicting scenes from the Hindu epic, the Mahabharata. Within the pages of this ancient text of

Hindu mythology there is a description of an nonphysical plane of existence in the universe called Akashic Records. Basically the idea behind the Akashic Record is that all information in the universe preexists and that individuals can actually tap into that information. The notion is that all information everywhere in the universe is somehow accessible. Some cultures actually say that this information is really a gift of the gods. That they left this behind for us to use at will. That we are able to download this information from other realms. This Hall of records, whatever you want to call it, is an interesting thing to correlate with Star Wars – the force be with you. Some kind of cosmic energy, omnipresence that we have the ability to tap into. This field of knowledge is something new that our own science is trying to discover and understand.

Astronomy was the precursor to modern day physics. The early philosophers were awed by the night sky and searched for logical models for the patterns of stars, the motions of planets, the rising and setting of the sun. Later, scientists like Sir Isaac Newton, turned their eyes towards the earth, questioning phenomena taking place such as - apples falling from trees, the flight of an arrow, the regular motion of a pendulum, winds, and tides. This questioning resulted in a set of "laws of physics."

The Renaissance period saw the development of a separate, distinct discipline – Physics. Our powers of observation sharpened over time, and with the invention of microscopes, telescopes, vacuum pumps, clocks, etc., more phenomena would be discovered that could be meticulously described by recording numbers in notebooks, by constructing tables and drawing graphs, and then by noting conformity to mathematical behavior.

Leonardo da Vinci is revered for his technological ingenuity. In his writings, da Vinci describes laying down on his bed, placing a candle at the foot of the bed and looking into the light of the flickering flame on the ceiling. This would put him into a very trance-like state, the state in which he would be able to imagine and come up with new ideas. The polymath conceptualized flying machines, a tank, concentrated solar power, an adding machine, and the double hull, also outlining a rudimentary theory of plate tectonics. Few of his many designs were ever constructed or even feasible during his lifetime; however, some of his smaller inventions, like an automated bobbin winder and a machine for testing the tensile strength of wire, entered the world of manufacturing without recognition of his contribution. Da Vinci made many important discoveries in anatomy, civil engineering, optics, and hydrodynamics, but because he did not publish his findings, they had no direct influence on modern science.

By the early twentieth century, atoms had become the frontier of physics with nuclei becoming the main focus of research by the forties. With each new development of instruments with ever-increasing power, scientists and researchers have been able to look more closely at things that are smaller and smaller.

In the 1900s, Serbian born inventor, Nikola Tesla pioneered modern electrical engineering and was the driving force behind today's wireless technology. In his writing, Tesla describes his method of approaching a complex engineering problem. Nikola Tesla was someone who wrote excessively about using very powerful visualization. In fact, he said that he was able to visualize a motor, then actually run research and develop (R&D) on the motor in his mind, so that he could tell were there might be mistakes or flaws. He would be able to do almost all of that in his mind, in this altered state of consciousness. Nikola Tesla is a person who believed in the existence of extraterrestrial beings. He really believed that he could communicate with other galaxies. Perhaps the greatest piece of evidence that Tesla believed he was in communications with otherworldly beings can be found in a letter he sent to the American Red Cross in December of 1900. In the letter he writes, "We have a message from another world unknown and remote. It reads (1) (2) (3).[4] Tesla was a genius with certain eccentricities. Everything he did was in threes. It is interesting, I think, because this some sort of symbolism. When we realize that so many of our ancestors, whenever they are talking about meditation with the divine and how divine downloads the information to them, we see this rich layer of symbolism there as well. Why was Tesla so obsessed with the number three? Could this be a clue that Tesla was using his own internal form of wireless communication to connect with the Cosmos?

India's remarkable genius Srinivasa Ramanujan was an unexceptional student; however, he is widely considered the greatest mathematician in history. According to his journals, Ramanujan believes he was being guided so cautiously by an otherworldly being, the Hindu god Namagiri. Basically everything that the family did was linked with this deity. Much of his creative work was actually done by going to bed with the intention to be visited by the Hindu deity that he worshiped. He would become lucid in the dream time and would receive enormous amounts of information. In one particular dream, he talks about seeing a very large red wall. A hand came out and wrote equation after equation. He memorized the equations and when he woke up he actually didn't want to work out the problems. Because he did his proofs on slate then transferred the results to notebook paper, it is rare to find the proofs. In fact, the sum total of the transferred note consists of only four notebooks. The vast majority of his equations were new and important and utterly groundbreaking. All of the science that we need to eventually be able to create portals, Star gates, travel teleportation. It all needs a mathematical foundation. And the closest that we have right now is revolution Ramanujan. The equations that we have been able to decipher so far are N-nn2=4a(u+7)-4n=8(mod32). It is an irrefutable fact that mathematics is the language of

[4] http://en.wikipedia.org/wiki/Nikola_Tesla

the universe. So if Ramanujan's formulas do in fact check out who knows what the potential is. It will be life altering. It could be revolutionary.

Albert Einstein would communicate with a realm outside our galaxy. A realm designed by God. Einstein would put himself into an altered state of consciousness which he called thought experiments – a kind of three-dimensional metaphoric reverie where he would come up with new ideas or visualize a problem. Einstein would put himself into a first person sitting on a beam of light traveling through space. What was the source behind Einstein's incredible insight about the laws of physics? Is it possible that Einstein's ability to enter an altered state of consciousness connected him to the inner workings of our universe? Einstein said that when he first developed the concept of relativity it was like a storm going off in his head. Albert Einstein was known to spend hours on end sitting in a chair formulating an idea or a thought to its inception. Was he tapping into some advanced yield of knowledge? Maybe he uses this type of technology to a certain level. In 1905, after immersing himself into these thought experiments, Einstein offers four papers that would refine mankind's perception of the universe in multiple ways. Included in that year was a paper which essentially proved that atoms were real entities that really existed. That was a very controversial idea at the time. In that same year he published a paper on what he called the fluorometric effect of light waves. Einstein said that light reacts with the particle and it comes in little discrete packets. One of his papers included the equation $E = MC^2$, stating that you could create and destroy mass. He gave us an idea of how mass might come into existence in the first place. The essay showed that mass can be converted into useful kinetic energy and so this is the basis of the new energy of nuclear fission. Another paper he published that year became the beginning of his theory of relativity. This was an amazing array of things to do for his science and entire career. The fact that Einstein did it all in one year is virtually miraculous. To get a sense of how dramatic those papers were, imagine that someone came up to you and said everything that you know about space and time and matter and energy is wrong. Then gave you all the math and all the explanations of why everything you know is wrong. It's often referred to as the Annus mirabilis paper, which basically means the miracle year or a wonderful year.

It could only be a coincidence that many of the greatest minds in human history have received information from an otherworldly source. Or could it be possible that Socrates, Leonardo da Vinci, Nikola Tesla, Srinivasa Ramanujan, Albert Einstein and other geniuses in human history used altered states of consciousness to access a realm that some people fear? Could they access the realm that contained the entire knowledge of the universe?

Look at history. Look at the people who are the Giants of Knowledge. Socrates understood what Einstein's papers say. This information wasn't theirs but something beyond them. Something larger. Something divine that allowed them to receive this information and give it to the world. All the knowledge of the universe already exists in the Cosmos. And maybe Einstein was able to access that knowledge because he was more in tune with the metaphysical world. This access into the metaphysical world may have been how he was able to unlock the secrets of the universe.

If so, where might this realm of knowledge exist? Perhaps answers can be found in Asia within the pages of an ancient Hindu text.

Is it possible that Albert Einstein's method of meditation, which he called his thought experiment, was actually connecting him to the Akashic Records? Could the Akashic Records be the source of inspiration behind the world geniuses? If so, how can such a metaphysical plane of knowledge exist? Perhaps the answer can be found in one of Albert Einstein's many theories.

In 1935 Albert Einstein published a paper describing the theory of quantum entanglement. It is theorized that if a particle is split into two parts both remain linked to one another, even when separated over vast distances. Experiments showed that two particles separated in space and time demonstrated a connectivity. If you should change the spin of one particle, the other particle would change. In other words, if something happens to one, the other responds in the same or an equal and opposite manner. This occurs instantaneously. So it's a form of instantaneous communication. It's what Einstein sometimes referred to as spooky communication at a distance. We have to remind ourselves that we are made of the Stardust. Our whole body consists of material that has been here since before the beginning of time. And these particles contain the knowledge of the universe. It's not only real, but it can explain to us so much weirdness in this world, which we still don't understand. This would include the idea of telepathic and remote viewing. These visionaries, according to their own diaries, believed they were in communication with some type of otherworldly entities. So were all these people just crazy, or is it possible that the universe is far more complex than we think? Is it possible that mankind's greatest minds had the ability to access the realm of knowledge through otherworldly beings? Could this be evidence that the Akashic records actually exist, just as the ancient Hindu believe? And, if a select few have the power to tap into this alien realm, does this mean it can influence them as well?

Albert Einstein's incredible legacy, which includes over 300 scientific papers, continues to influence the world more than a half century since his death. And while Einstein's theories define our understanding of the universe, perhaps his greatest contribution to mankind is still to come. Time travel. Einstein's special theory of relativity and the speed of light allows for the real possibilities of time travel. Some scientists today believe that Einstein's theory of relativity creates the possibilities to travel through a portal or worm hole called an Einstein Rosen bridge. A space anomaly that may actually be a gateway to a parallel universe. In the theory of general relativity, space actually acts like something that you can bend, warp and make holes in it. An Einstein Rosen bridge is a particular example of a wormhole. A wormhole is a structure in space and time that involves kind of tearing a hole in space. Stretching the material, if you think of it being a material and reconnecting it somewhere else. Einstein Rosen bridge takes two points in space and connects them so that if you travel at one end you come out of other. And even though those two points appear in different universes, it's essentially a teleporter, something that allows you to get between two different places instantaneously. Could Einstein's theory article actually serve as a portal to another universe, perhaps opening a gateway that will one day establish contact between humans and extraterrestrials? Or could it be that Rolando's spherical edged universe makes it unnecessary to bend or tear that space?

Maybe there is a secret code on earth and our universe. A secret code that needs to be deciphered by a future generation that understands what really happened in the past and present. One that will understand the future. That generation is us.

Do the world energy grid, ancient star alignments, ancient flightpath, and mathematics represent some sort of code? Is there a secret code embedded in the stone monuments of early man? One that reveals the truth, if so why? Was it to test us to see if we could figure out the code for ourselves, or is it simply to communicate with us? Is there a connection between the multi-universe? A mathematical one? Is there no explanation? Maybe it is just that way because it simply is. Or is it really God? A higher power exerting it's presence into our thinking?

THE BIG BANG
Experts say the Big Bang universal intelligence started from nothing at all. No God.

PARALLEL MULTIVERSE
No mathematical connection to our universe, no God

BUBBLE MULTIVERSE
No mathematical connection to our universe, no God

TREE MULTIVERSE
No mathematical connection to our universe, no God

SOCRATES
The father of modern western philosophy, HEARS in his head DEMON, ANGELS, MAYBE A GOD

PLATO
Believes in a mathematical universe sphere, mathematical earth sphere, 12 faces, 20 points

LEONARDO DA VINCI
Put a candle at the foot of his bed. Flickering light on ceiling came up with new ideas. The father of modern world artist. May be a god

ISSAC NEWTON
Key figure in the Scientific Revolution. Gave us the first practical reflecting telescope. In addition to his work on the mathematical sciences, Newton dedicated much of his time to the study of biblical chronology and alchemy

NIKOLA TESLA
Driving force for wireless technology. Visionary. We have a message from another world and remote 1 2 3

SRINIVASA RAMANUJAN
Greatest mathematician in the history of the world. While asleep Hindu god. N-nn2=4a(u+7)-4n=8(mod32)

ALBERT EINSTEIN
The most intelligent man in history, UNIVERSAL INTELLIGENT ORDER, altered state of THINKING consciousness, maybe God

ANGKOR CAMBODIA
Hindu Temple. Metaphysical Universe

AKASHIC RECORDS
Universal Information, Individuals can tap into.

All of these famous people had some sort of otherworldly help. Now Rolando Rivero's conversations with God do not seem so crazy after all.

ROLANDO RIVERO
12 + 20 + 1 = 33 $O = O^{12}\ \Delta^{20}$

Rolando's creation code formula has a direct impact on these unfinished codes that I helped finish as a part of a set. Quantum mechanics.

PLATO
12 FACES 20 POINTS 12 + 20 + 1 = 33

NIKOLA TESLA
1 2 3 1 + 32 = 33

SRINIVASA RAMANUJAN
8 (MOD 32) 1 + 32 = 33

Creation Code

Rolando Rivero

$$O = O^{12}\ \Delta^{20}$$

Multiverse
12 + 20 + 1 = 33

Sugar DNA 8+8+8+8=32=1 double helix

Silicon Quartz Crystals 32 + 1 = 33

Bee chromosomes 32 +1=33

Rolando Rivero's creation codes all have a direct connection and will be explained as you read on.

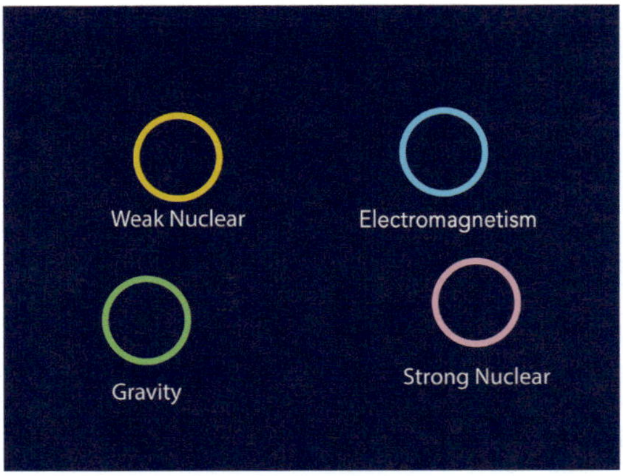

Before the Big Bang

According to scientific experts and physicists, our Big Bang was created of nothing at all. This seems to be scientific, but doesn't quite answer the question of how the universe began. They say that 14 billion years ago there was a point—a very, very small point—this small point is smaller than an atom—so small they can't measure it. But they also say that just before the Big Bang there had to have been four forces—gravity, electromagnetism, weak nuclear, and strong nuclear. Somehow gravity detached from the three other forces and this caused the Big Bang. Then in less than one second, our universe expanded way too quickly, expanded faster than the speed of light. It didn't make any sense. According to the laws of physics there is nothing faster than the speed of light.

So they created a timed distance called planks time. So a billionth, billionth, billionth, billionth, billionth, billionth of a second inflation expanded. Then, gravity took over again. The premise of me believing that our universe started from practically nothing over 1 billion galaxies, hundreds of billions of stars, our solar system, Earth—everything in our universe. That is a lot of stuff that began from nothing. And that is the reason why these experts created inflation theory. If you tell me maybe a planet or two, okay; our solar system, our Milky Way, galaxy; maybe that's a little more believable. Also, gravity plays one of the most important roles in our universe. You just can't turn gravity on and then gravity off like a switch. That's pretty hard for me to swallow.

Also scientific experts and physicists say our universe is expanding too fast. This is mathematically impossible, yet they have data information to prove it. They say some weird and strange force, like dark energy, is forcing everything in our universe to expand. That perhaps one day everything you see in the universe would expand so far away from each other that everything would turn off and not be a part of anything. That everything would just die out.

Our universe is flat and infinite, there is no edge or other side. So where does dark energy and dark matter come from? No one knows. In the beginning there was darkness and then a bang, giving birth to an endless expanding existence of time, space and matter. Now we see further than we ever imagined – beyond the limits of our existence into a place we call the universe. Maybe there are other worlds out there; universes germinating off other universes where there could be an exact copy of our solar system on earth and each one of us.

Where are these other universes? We can't see them because we are in different dimensions. The question isn't whether they exist, but whether there is one, two, three or four different kinds of universes. Our universe sits in a sea of parallel universes. Imagine another world, a whole other universe, with a solar system and a planet exactly like ours. On this parallel Earth there lives an exact copy of you, who could be leading exactly the same life but in another universe.

Now imagine an entirely separate universe. We could be living in a slightly different life at the same time as you live this life. It may seem fantastic, incredible and impossible, but there could be many other universes out there and now some of the world's leading physicists believe that they have evidence to prove it. We are experiencing an extensional shock.

Our world views have been shattered with the realization that yes, there could be parallel universes. One of the real exciting things is that scientists today are trying to see whether parallel universes are a physical reality. Amazing new discoveries about the true size of the universe seem to indicate that all that we see is not all that there is. Remarkably there may be four types of parallel universe out there. One kind could exist in the same exact space we're in, but it is so far away that we cannot see it or reach it. In another scenario multiple other universes could be in a giant cosmic soap bubble adrift in a cosmic sea of giant bubbles.

Revolutionary developments have changed the entire landscape. Data from our space has given us a new look at cosmology and satellite data indicates there could be parallel universes. In yet another scenario many parallel universes occupy the same space and time as our universe but because they are in different dimensions they are invisible.

In yet another, all the laws of physics are different so everything looks completely different. New theories called string theories are giving us worlds of higher dimensions. Quantum physics at the microscopic scale is also revealing to us the fact that there could be parallel universes. To keep things simple physicists have divided parallel universes into different categories.

Rolando Rivero's Exit Side Scientific Proof – *In 2010, Roger Penrose, known for his work in mathematical physics, reported possible evidence, based on concentric circles found in WMAP (NASA's Wilkinson Microwave Anisotropy Probe) data of the CMB sky, of an earlier universe existing before the Big Bang of our own present universe. According to this controversial study, ring-like patterns in the sky are thought to be ghostly imprints of a universe that existed before the big bang. If that data proves correct, these cosmic rings may provide actual evidence that our current universe is just the latest in an endless string of recycled universes.*

Ingunn Wehus and physicist Hans Eriksen, both from University of Oslo, redid Penrose's analysis of data and an independent group led by University of British Columbia's Adam

Moss made a similar analysis, and published their results in the Journal of Cosmology and Astroparticle Physics April 26, 2011.

Incredibly, both groups saw the same circles that Penrose did. They say they have discovered ***12 concentric circles****, some of which have five rings – which means the same object has had five massive events in its history. The data also showed concentric regions of low variance in the shape of* ***equilateral triangles. The circles and triangles are really there****, confirming that my theory has merit. Proof that what follows explains the presence of our multiverse and how it began.*

Creation Code

Rolando Rivero

$$O = O^{12} \Delta^{20}$$

Multiverse
12 + 20 + 1 = 33

Sugar DNA 8+8+8+8=32=1 double helix

Silicon Quartz Crystals 32 + 1 = 33

Bee chromosomes 32 +1=33

Rolando Rivero's Big Bang

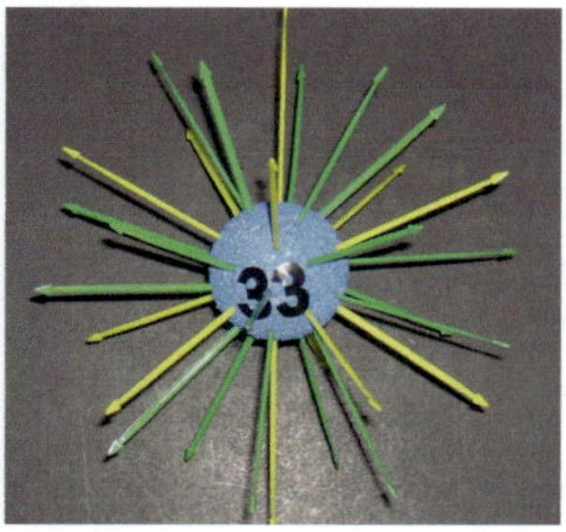

The exit side of the universe has 32 direct multiverse. These 32 multiverse are the older siblings of our universe and cover our universe. Our universe is in the center. Trying to identify for mapping simplicity, I have colored our universe blue with the number 33. Also as verification in a bigger scale (ZONE ONE). So in the real beginning of our universe, before the Big Bang, our universe is a pressurized dark void, the 32 multiverse gave our universe it's finite spherical shape and it has an edge. In the quantum realm, almost simultaneously, these massive 32 multiverse, 20 triangular light pulsating, vibrating string energy points and 12 circular, vibrating string energy points had begun penetrating our universe's edge. And gravity and dark energy was born pushing dark matter aside in our universe. The fight begins between gravity, light energy, dark matter, and dark energy. You see, gravity was already present before the three other forces, electromagnetism, strong nuclear and weak nuclear appeared to enable the Big Bang. These massive 32 multiverse

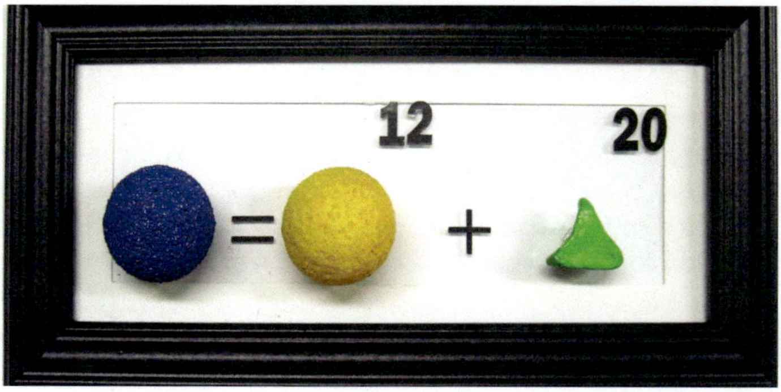

will feed our universe to the center to one single point of our universe. Imagine, at the edge of our universe, these massive, really massive 20 triangular energy points in motion and 12 circles rotating all positioned timing like a clock, compressed to the center of our universe to this one point. This one point is far from being small.

On the contrary, this is–Albert Einstein E= MC2 on steroids: 20 + 12 + 1 = 33 or 1 = 20, 12. Unprecedented perfection and the Big Bang number 33 multiverse was created. Because this force was so massive our universe expanded really quickly, it overwhelmed gravity. But gravity did not turn off, it hung on for the ride. Also something else happened. Quantum dark energy now has a twin brother, dark energy. Quantum dark energy and dark energy strangely were born at a different age and time but they go hand in hand.

Consider this:
Quantum string energies before the multiverse big bang
Quantum String light energy + dark matter = Quantum dark energy

Energies after 33 multiverse Big Bang
String energy + dark matter = dark energy

Twins are always connected to the quantum realm and reality. So in other words, Quantum dark energy = dark energy. They are one in the same, yet different. Remember Einstein's "spooky communication at a distance" theory discussed in his quantum entanglement experiments?

This inflation expanded our universe really fast and vast. Our big bang, young and energetic did not realize the quantum older sibling 32 multiverse still had control and quickly slowed down expansion and inflation of our universe. Fourteen billion years ago, our big bang was created and everything in it from all particle elements, clusters of galaxies, billions of galaxies, hundreds of billions of stars, our milky way, our solar system, Earth and us. God created 32 multiverse and everything in it unprecedented perfection with intelligence and order to run like a clock. Everything also has to have a balance, an opposite attraction. Cause and effect. For example 32 universe to create one universe, quantum physics and physics, life and death, white and black, good and evil, light energy and dark energy, just enough imbalance to keep the realm of change for recycle to never stop. God designed it that way. The 32 multiverse pulsating vibrating string energies created our beginning, our present, our future and today it is expanding, fanning and keeping our universe in motion. Until we have expanded to our full capacity, because we do have an edge.

Typically, multi-verse theories describe hypothetical sets of multiple possible universes that together comprise everything that exists or can exist. This includes the entirety of space, time, matter, and energy as well as the physical laws and constants that describe them. Outside energy sources are described but not speculated into detail as to what is beyond the geometrical form of the multi-verse. The exact definition of these parameters is not explained.

This multi-verse theory has much stricter geometrical conditions in regards to the relationship of the universes within the multi-verse. Some theories describe endless parallel universes and others portray random bubble universes. In this theory all of the universes other than ours have a very direct energy impact to ours based on their surrounding locations. This theory also clearly renders our universe to be in the center of the multi-verse as opposed to equal or parallel.

Rolando Rivero's multi-verse theory clearly indicates that the universe, though seemingly infinite, only has a certain maximum volume to expand itself within. The universe as we know it is the same overall size as all of the other universes. Though our universe is the youngest and still expanding within its confines, the overall dimensions are the same. This basically means that although the universe may not have fully filled its container yet, its spherical containment is the same size as the other older universes. Where other universes have already expanded to their maximum size within their holding units ours has vast empty space between the furthest outer points of the universe's expansion and the hypothetical walls of the universe we are within.

During the inflation and expansion period everything was condensed into a very small area—smaller than a single atom.

The perfect amount of gravity, a big bang—an incredible explosion— scattered everything all directions within less than a billionth of a billionth of a billionth a billionth of a billionth of a billionth of a second. Expanding faster that the speed of light. A speed so fast that science had to create a new unit of measuring time = Planck time. It is the time required for light to travel, in a vacuum, a distance of 1 Planck length and it is named after Max Planck, who was the first to propose it.

Expansion in still fanning and causing the universe to expand and move, colliding making more motion and movement.

Is it possible for over expansion to occur? Would everything fall apart? Or will the universe expand so far apart that we can't see it anymore? Or will it cause a big crunch? Theory?

What is the glue holding all the galaxies in place? Is it making it move apart or together? Or are we being swallowed into another universe?

CONTRASTS BETWEEN MULTI-VERSE THEORIES

Most theories speculate as to how many possible universes there could be within multi-verse theories. Rolando Rivero's theory specifically defines the multi-verse to be a 33 universe system, with ours being in the center surrounded by 32 other separate universes.

There are areas where the surfaces of the universes make contact with one another. There are also travel zones within the space between the universes where they do not make direct contact.

This theory also details how the other universes caused the big bang which created this universe.

This theory says the first universes within the multi-verse theory were the outer most ones which filled certain pockets and then sprung forth the inner universes. As these 32 other universes existed geometrically there became a central focal point of the gravitation of their energies towards the center of the multi-verse. This consistent energy focal point was pressurized over time and eventually became ground zero for the big bang incident.

This theory indicates that the overall shape of each universe is a perfect sphere as opposed to other theories' predictions that our universe is of a shape more closely resembling a disk.

This theory is strictly a definition of the overall geometry of the multi-verse and the space between the universes. This theory does not attempt to explain the mathematical ratios or the individual behavior of the universes within the confines of their unique systems.

In this multi-verse model, eternal inflation is not a true mathematical possibility as the universe, endless it may seem, does have finite dimensions.

This multi-verse model satisfies the condition that from the exit-side of the multi-universe, you cannot directly view any portion of the 33rd central universe, as it is completely covered and surrounded by its older siblings.

This is where everything started. Before this was a really big dark pressurized void and the beginning of God's pulsating energy. The beginning of gravity, dark matter, dark energy, the speed of darkness, which is faster than the speed of light. Because of darkness light was created so that light could travel. All of which is followed by the big bang.

The current scientific theories regarding the beginning of the universe claim that it was started spontaneously from nothing. When put into simple terms, these theories sound very scientific. The objective of this multi-verse theory is to present a logical explanation and geometric model for the development and currently existing state of the universe and its counterparts.

Before the overview of the multi-verse is presented, the model and its components must be explained. In order to fully comprehend what the overall representation shows one must understand what the pieces exemplify.

The overall shape of the universe is continuously being debated due to the limited amount of evidence obtainable from the Earth's position within the system. Throughout the explanations and models of this multi-verse theory, the overall shape of all existing universes will be represented as spherical (Fig. 1), regardless of the age and content of each, as a means to move the discussion towards a larger model of a multi-verse.

- *Figure 1. The assumed shape of any universe for the sake of discussion. This multi-verse theory categorizes all of the existing universes into three zones. Zone One is the centric location in the multi-verse which the known universe inhabits. Zone Two encompasses all universes directly adjacent to the known universe. Zone Three groups together all universes that are not bordering the known universe.*

In all models pictured: The known universe within Zone One is represented as a blue sphere, all universes within Zone Two are displayed as yellow spheres, and all universes within Zone Three are shown as green spheres (Fig. 2).

- *Figure 2. This is a simple model to illustrate the three different zones. The known universe is shown as the blue sphere in the center, the neighboring universes are portrayed as yellow spheres, and any universe not directly bordering the known universe is depicted by a green sphere. This is not a complete model of the multi-verse. It is merely a portion of the multi-verse that has been isolated in order to demonstrate the definitions of the different zones.*

Explaining the origin of the known universe is a top priority for scientists. This has always seemed difficult to accomplish simply because many scientists go about the process incorrectly. Most theories explicate the origin of the universe within the confines of an individual system, limiting the resources available to create the needed effects. The easiest way to describe how the universe was formed is by first clarifying the theory of the multi-verse as a whole. Once the complete shape of the multi-verse is defined, it is then possible to designate the derivation of energy sources and other factors contributing to the birth of the known universe without them being considered external influences. Therefore, the geometry of the comprehensive multi-verse must be identified prior to detailing the formation of the known universe.

Patterns are prevalent within the known universe on any level of examination. Geometric configurations are easily observed in atomic theory and other microscopic studies; it should be of no surprise then, if the overall shape of a multi-verse is methodically formatted. Seeing as the universe is systematically structured on its smallest level, theorizing that it has predominant geometric features on a larger scale is a valid hypothesis. As the multi-verse model is explained, it can be noted that any and all portions of the model are seen to have arithmetical correlation.

In Figure 2, slim cylindrical connections between the spheres can be noticed. These lines serve two purposes. The first reason for the existence of this system of links is very simple and unrelated to the actual multi-verse theory in that they simply hold the physical illustration of the model together. They coincidentally also serve a representative purpose for the multi-verse theory by portraying roughly estimated paths of string energy.

All of the universes within the multi-verse have an effect on one another through the energy they naturally produce. The energy emitted by Zone Two universes are not visible within the portion of the model shown in Figure 2. The only string energy diagrammed is that which reaches the single universe within Zone One from the other universes within the multi-verse. Zone Two universes are directly adjacent to the Zone One universe and do not leave room within the model to show the energy relation. If a model were to show the relation between a single universe in Zone Two to that of the Zone One universe or specifically display any space that may exist between them, then string theory could be exhibited between the two members of the system.

This multi-verse theory does not claim that the universes within the multi-verse only expel string energy in one direction. The only string energy demonstrated is that which is directly relevant to the Zone One universe.

Figure 3 shows the Zone One universe in a separate diagram to better expose these specific paths of string energy.

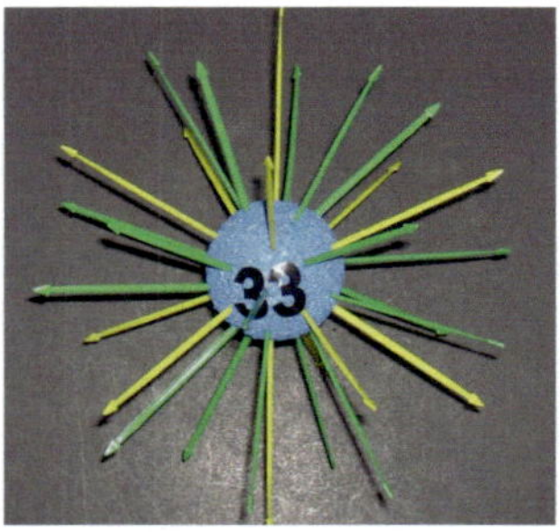

- *Figure 3. The paths of string energy from Zone Two universes are shown as yellow connections and those from Zone Three universes are shown as green links.*

Figure 3 also shows the Zone One universe being branded as "33" for diagramming purposes. The universes within the multi-verse are given identifying numbers in order to help visually map the multi-verse. The details of and reasoning behind specific functions of this nomenclature will be further explained shortly.

The multi-verse system defined within this theory contains thirty-three separate universes. Zone One contains the single known universe,

Zone Two encompasses the twelve adjacent universes, and Zone Three comprises twenty outer universes.

Due to the symmetry of the geometric form of the multi-verse, the designation system is partially relevant on three separate levels. The outer universes of Zone Three are labeled as Universe 1 through Universe 20. Specifically which of these are labeled as which numbers is irrelevant, however, as long as they are given unique aliases in between one and twenty. The universes within Zone Two are of similar manner in that they are defined as Universe 21 through Universe 32 but which universe is labeled which number is not necessarily of the highest importance. Zone Three only holds one universe, so its numeration does not need to be debated and is labeled as Universe 33.

Each universe is given an identification number on the model simply to assist visual comprehension of the representation. Physically labeling each sphere also helps to characterize the symmetry of the system by allowing the spectator to realize that they are viewing the multi-verse model from different angles. If the spheres were not labeled, an audience would be incapable of determining which segments of the multi-verse were being displayed in a picture. This cataloging is necessary since there are limited resources when showing a three-dimensional model via a two-dimensional communication medium.

Rolando's Theory of How the 33 Multiverse Began

First, God's light in the center of Heaven opened space for darkness, stretched and stretched, until inner Heaven seemed like a compressed sponge. There is this really, really big dark void. Extremely huge. This massive dark void is far from being empty. This void is a sphere with its own edge. So in the real beginning God's energy was coming out from every angle. But this was a pulsating string energy with patterns. Intelligent mathematical patterns. And this massive void has an extreme amount of pressure. There is no name for this pressure. I am going to call it Rolando's Dark Pressurized Void. So as soon as God's energy penetrated this void, gravity and dark energy came into existence. For a very, very long time these forces were in a tug-o-war battle pushing and pulling. At first it seemed that the pressure of the void and the dark energy were winning because it contained God's energy at the outer edges. But this was God's design.

Creation Code

Rolando Rivero

$$O = O^{12}\ \Delta^{20}$$

Multiverse
12 + 20 + 1 = 33

Sugar DNA 8+8+8+8=32=1 double helix

Silicon Quartz Crystals 32 + 1 = 33

Bee chromosomes 32 +1=33

Because of the energies colliding on the edges, a curtain of elements was formed. These elements created cosmic rays of light and heat for a long time. Secondly, because of this, certain areas of the outer edges were changing. Some areas were soft while some were hard. A gassy, milky, silicon crystal. And the patterns were consistent. Time would pass and now in a lot of areas triangular shapes, tetrahedrons were formed. And in these tetrahedrons, elements were constantly pushing and pulling hand in hand tug-o-war battle. Dark pressure void with dark energy, gravity giving a helping hand. These elements were in motion to these tetrahedrons and bouncing off, creating craters in circular motion. And between the voids' pressure, the first 20 green multiverses were born and at that time 12 yellow multiverses were taking shape. But these 12 yellow multiverses had smaller craters on edge and more help from the green multiverses. So between small void edge craters, green multiverses and pressurized void the inner sphere formed 12 yellow multiverses. And the full expansion of 20 green multiverse and the full expansion of 12 yellow multiverses – 32 pulsating string energies all together entered into the center of the one pressurized dark void. Each of the multiverses would have their own four forces – Weak Nuclear, Electromagnetism, Gravity and Strong Nuclear plus similar to our universe their own big bang, inflation and expansion. In the center point of this dark void, all 32 pulsating string energies collided to become our Big Bang creation. The beginning of our universe.

There is no way possible to recreate the Big Bang or the beginning of life because we cannot know the exact "Goldilocks" ingredients or the environment at that time. All is hypothesis. Recent scientific thought suggests that amino acids from comets bombarding the Earth may have been responsible for the beginning of life.

Michau Kaku suggests that comets do give us clues as to how the Universe was formed. They date back to the birth of our solar system some 4 1/2 billion years ago. The theory goes that giant clouds of dust and gas floating in the atmosphere fused with ice crystals growing into larger masses. Some of these masses formed planets. Others are still floating around in space. Scientists believe that the comets are leftovers from the forming of our solar system and that they remain unchanged from that time period. They believe that if you could crack open a comet you would discover what elements existed and how they contributed to the development of life on the planet.

Beyond the planet Neptune lies the Kuiper Belt, a region of space filled with such remnants of the early solar system. There is a similar, but less icy, asteroid belt between the planets Mars and Jupiter. Within the Kuiper Belt are many dwarf planets, worlds too small to qualify as planets yet too large to be called asteroids. The planet Pluto was the very first Kuiper Belt Object to be seen, but it wasn't until 1992 that scientists discovered the slow moving world in the outer solar system. Sedna (sed'nah), about three-fourths the size of Pluto, was discovered in 2004. It is so far out from the sun it takes about 10,500 years to make a single orbit. In 2005, astronomers discovered an object in the Kuiper Belt thought to be larger than Pluto, though further research showed that it was slightly smaller. Known as Eris, it orbits the sun approximately once every 580 years, traveling almost one hundred times farther from the sun than Earth does. This discovery led to some astronomers having a problem with terming Pluto a full-scale planet, so in 2006, Pluto, Eris, and the largest asteroid Ceres were reclassified as dwarf planets. Two more dwarf planets, Haumea and Makemake, were discovered in the Kuiper Belt in 2008.

Our known Solar System is divided into three parts: rocky planets like Earth, which are close to the Sun; gas giant planets, which are further out; and frozen objects within the Kuiper belt, which lie just beyond Neptune's orbit. Beyond this, there appeared to be an edge to our Solar System where only one object, Sedna, was previously known to exist for its entire orbit. However, Scott Sheppard and Chadwick Trujillo of the Gemini Observatory now say there is a very distant dwarf planet they are calling 2012 VP113, that is even further out. And they further believe that there is an even larger planet beyond that could potentially influence 2012 VP113's orbit as well as other objects within the Oort cloud.

With further research and advancing technology expanding what we can see in space we may surely discover more and more worlds that had not been previously known to us.

Over Quadrillion of Space Size

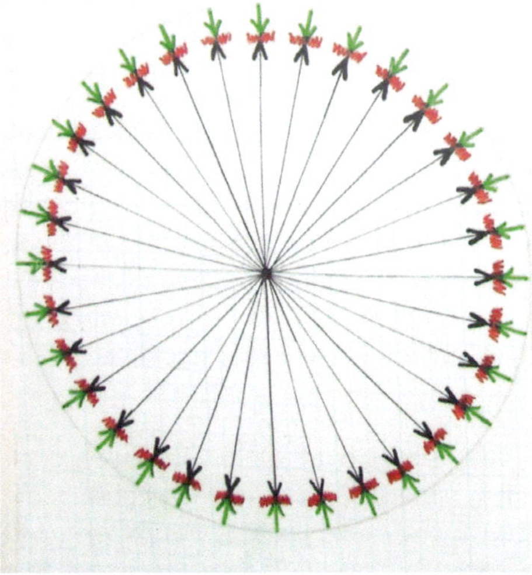

All around this sphere pressured dark void God's light energy began pouring into it in all directions. Here is where gravity also first began. Let me explain, physicists and scientists have a hard time. They cannot really explain how the Big Bang and gravity began or how it works. It's hard to pinpoint a mathematical solution when there is no beginning and no end. By having an edge and this quantum pressured dark matter makes it simple. If not, all of our billions of galaxies and hundreds of billions of stars in our galaxy, our solar system, our Earth would just fall. Look at it like a really big water balloon and everything in the universe is inside. The balloon would have an edge with its own maximum compartment and all matter would be occupied in space. Similar to our universe with hydrogen, if the water balloon has a hole, the water would be coming out because there is pressure confined inside.

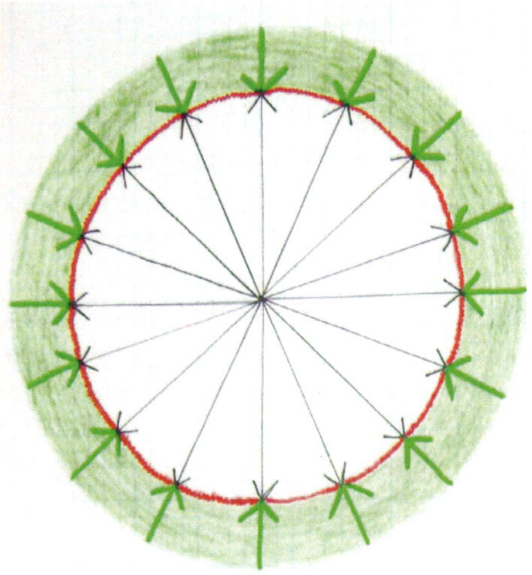

Let's continue. So at the same instant dark energy was also created. Between light energy and dark matter, dark energy was created. How was dark energy created? Think about Albert Einstein E=MC^2, but in the quantum world you have two matters in motion, which are light energy and dark matter colliding, crashing, smashing and in a constant arm pulling back and forth. This is how dark energy was created.

Silicon like milky cloudy film

Triangles are creation helpers. Crystallizes from pressure and has magnetic quality

Formation of green universes

Now dark matter having the upper hand was winning the fight pushing light energy and dark energy right back but God has it all planned out and released more and more light energy but with intelligence. These energies are now pulsating and vibrating and string energy was created. So while these energies are crashing and smashing into each other, they are creating heat, light energy and dark energy, more organized, bouncing off the edge of this massive pressurized dark void. Three more forces, electromagnetism, strong nuclear, and weak nuclear had emerged. According to scientists in our universe, these forces are how our universe was created. And nothing existed before. Everything was created from nothing. All new particles are starting to emerge like the particles in our universe. The two important particles at this time are hydrogen and silicon. A new substance is starting to emerge.

And God said, Let there be a firmament in the midst of the waters, and let it divide the waters from the waters. And God made the firmament, and divided the waters which were under the firmament from the waters which were above the firmament: and it was so. And God called the firmament Heaven. And the evening and the morning were the second day. Genesis 1:6-8

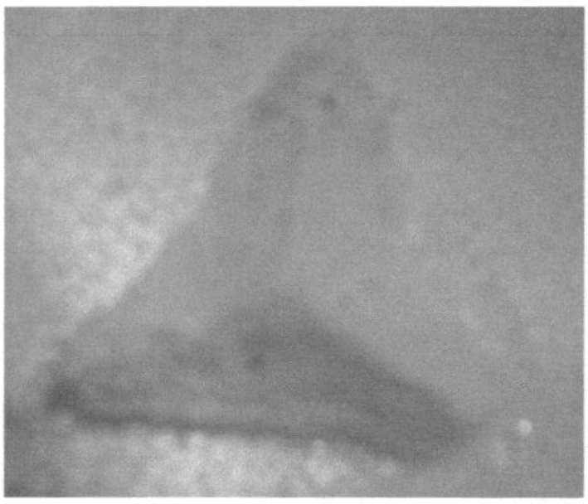

This is where everything started. Before this was really big dark pressurized void and the beginning of God's pulsating energy. The beginning of gravity, dark matter, dark energy, the speed of darkness, which is faster than the speed of light. Because of darkness, light was created so that light could travel. All of which is followed by the big bang.

Hydrogen is the most abundant particle in our universe or multiverse. So between hydrogen being water and the heat from constant cosmic burst created a gassy, milky-like substance called silicon. Silicon is the eighth most abundant particle in our universe. Perhaps billions of years of arm pulling of all these energies created a gassy milky silicon triangular shape, all around on the edge of the dark pressurized void.

Because these triangles have three sides and are slightly curved, these energies were bouncing off these silicone triangle curved lines, creating a circular motion. Perhaps billions of years in the making these gassy milky silicon triangles hardened to be silicon crystal triangles, with a more defined circular rotation, really big massive craters were created. The continuous motion were all of intelligent circular rotation. Precise architecture from God created 20 perfect multiverses. Since these 20 multiverses are closer to the edge of the void, and between the void pressure dark matter and God, light energy was really strong and powerful, still feeding into the center of these 20 multiverses. The governed four forces, gravity, electromagnetism, strong nuclear, weak nuclear had their own Big Bang. Unlike our universe, our Big Bang, scientists and physicists say our universe, our Big Bang, started from nothing at all.

Circular energy

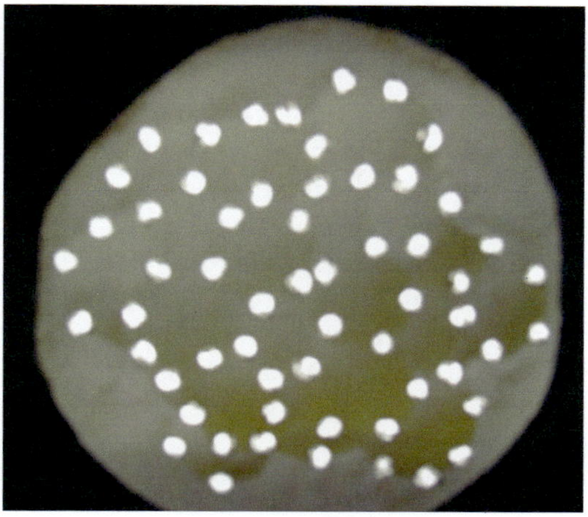

Triangular points God's light energy

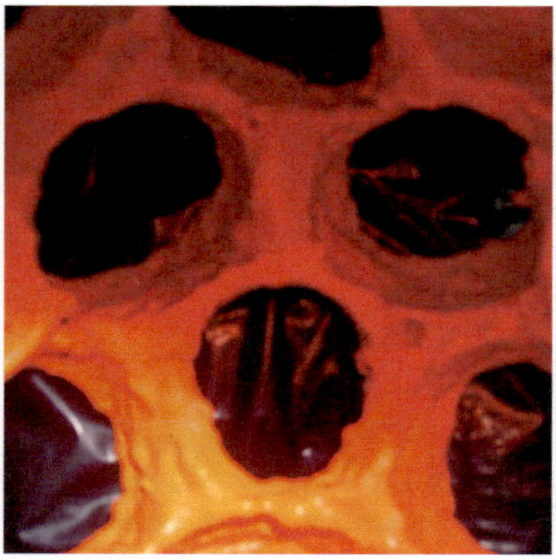

Inside of universe looking out as it is being filled up – neutral areas traveling between universes (wormholes)

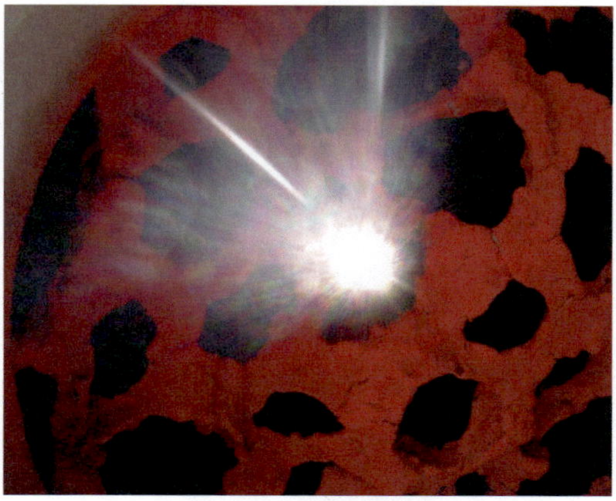

Light coming into universe

Multiverse dissected in half. Light shows through the triangular areas

It is he that sitteth upon the circle of the earth, and the inhabitants thereof are as grasshoppers; that stretcheth out the heavens as a curtain, and spreadeth them out as a tent to dwell in: Isaiah 40:22

The phrase "stretcheth out the heavens as a curtain" may be a reference to the expanding universe, as modern astronomers think of it. The Holy Scripture references the "stretching-out," or "spreading-out" of the heavens (space), when God created the universe many times. Look at, for example, Job 9:8; Psalm 104:2; Isaiah 42:5; 44:24; 51:13; Jeremiah 10:12. Or the "heavens" as referred to here may simply be referring to the atmospheric heavens, spread out like a curtain, or "tent to dwell in," around the circle of the earth. This atmospheric "tent," refracting and spreading light over the hemisphere, is sharply distinct from the darkness outside.

It is beyond all the stars and galaxies and seemingly has no end.

Third Zone Green Universe Total = 20

First, in the beginning of the 20 multiverse, almost simultaneously, 20 spheres were born on the outer edge of Rolando Rivero's dark void. Note: in trying to identify for mapping simplicity, since there is no research for mapping multiverses, I have colored these 20 multiverse with a green color with numbers 1 through 20 also as verification in a bigger scale (ZONE THREE). All of these 20 multiverse have the same size, shape and characteristics of our universe. Without getting too complicated, for example, each one had its own four forces – gravity, electromagnetism, strong nuclear, and weak nuclear, Big Bang, all particles, full expansion, billions of galaxies, hundreds of billions of stars, and planets. Unlike our universe, which is still expanding today and has not yet reached its full compartment, these 20 multiverse have expanded to their full capacities and helped with the creation of 12 inner multiverse.

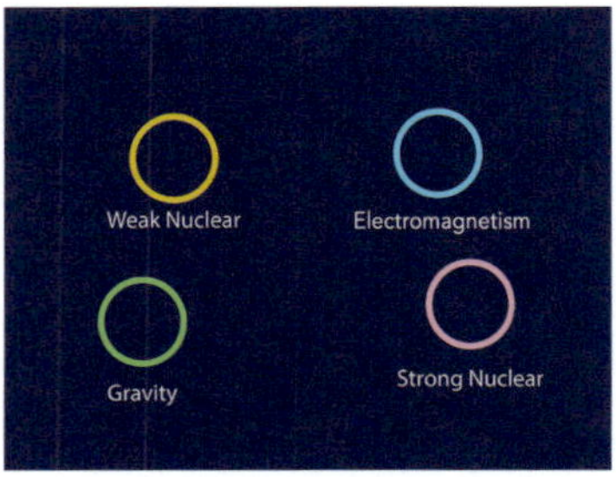

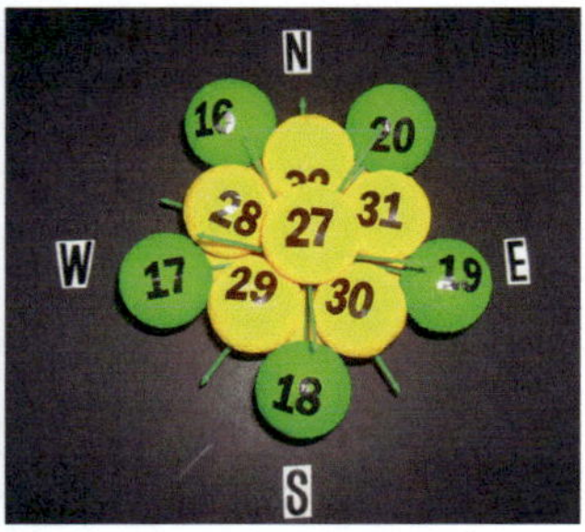

Green multiverse front, back side.
Mapping 20, 19, 18,17,16.

Green multi-verse front side mapping 15, 14, 13, 12, 11 (Note: 27 yellow is purposely left out to view center blue universe #33)

Close Up View of Green Multiverse, Yellow Multiverse and Blue Universe

Close Up View of Green Multiverse, Yellow Multiverse with Yellow number 27 covering over number 33 Blue Universe

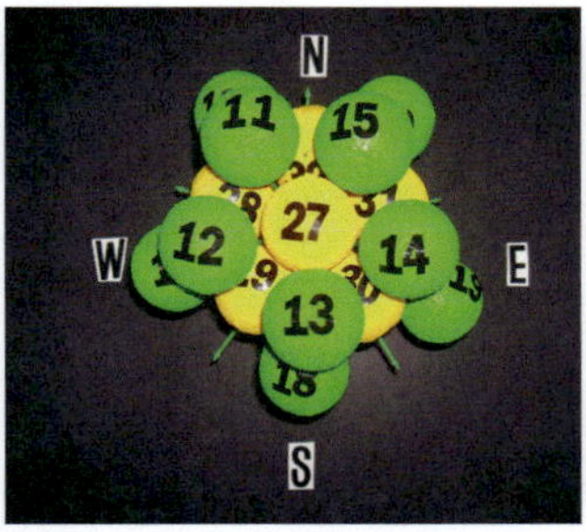

Complete mapping of front, back side of Green multiverse

West View mapping green multiverse

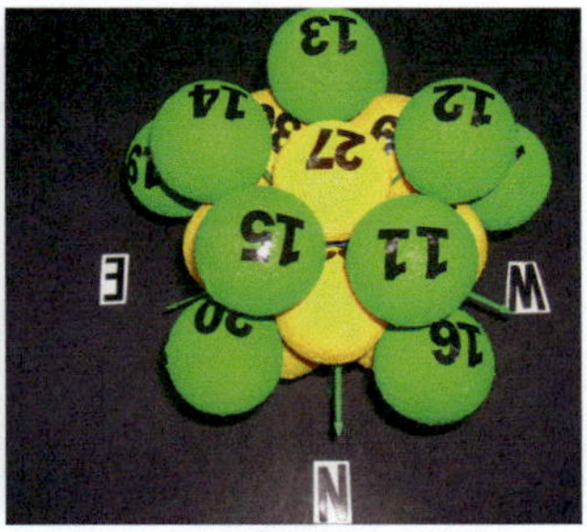

North View

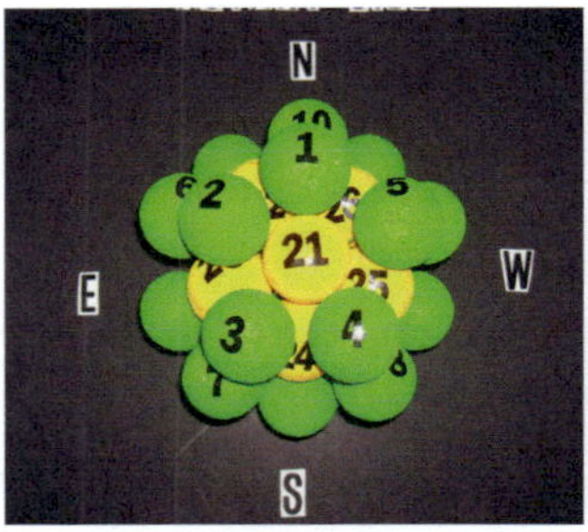

Green multiverse rotary side back 10, 9, 8, 7, 6 (Front side 5, 4, 3, 2, 1)

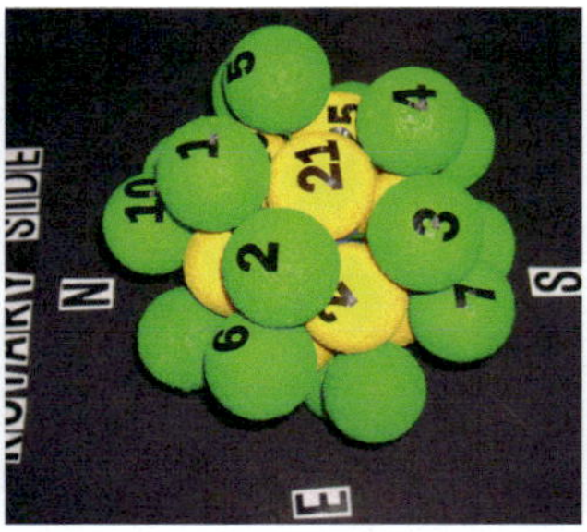

View rotary side East

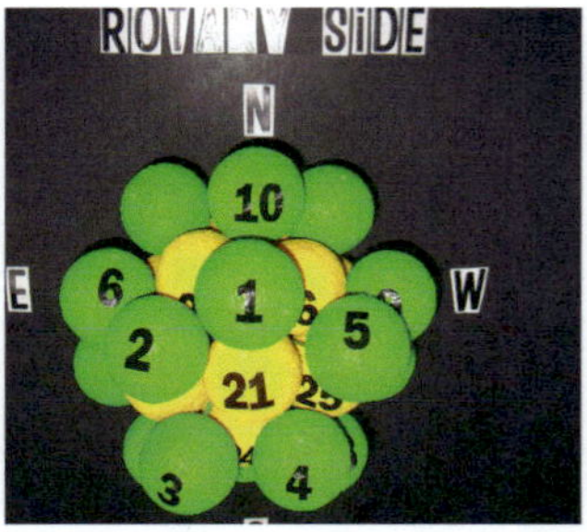

View rotary side North

East View

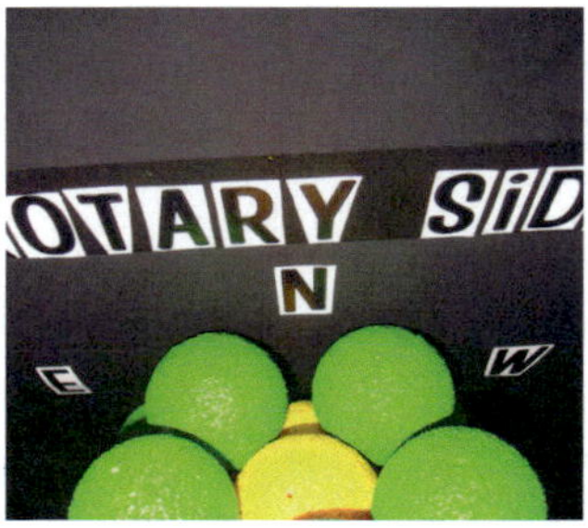

View Rotary side close up North

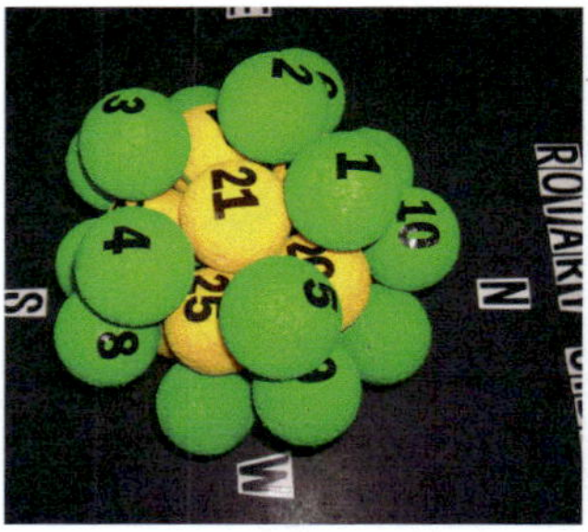

View Rotary Side West

View Rotary side South Close Up

View of South West

View close up of West side.

View rotary close up front

Note: without numbers geometric would all look the same

Thou, even thou, art LORD alone; thou hast made heaven, the heaven of heavens, with all their host, the earth, and all things that are therein, the seas, and all that is therein, and thou preservest them all; and the host of heaven worshippeth thee. *Nehemiah 9:6*

Second Zone Yellow Universe Total = 12

Similar to the green zone multiverse, these 12 multiverse began with smaller silicon crystal triangles but with smaller craters at the edge of the dark void. More help from the direct touching points of the older 20 green zone multiverse siblings. So between God's energy, dark energy, silicon crystal triangles, smaller craters, pressurized dark matter void, and the 20 green multiverse, 12 inner multiverse were created. Note again trying to identify for mapping simplicity, since there is no research for mapping multiverses, I have colored these 12 multiverse yellow with numbers 21 through 32 also as verification in a bigger scale (ZONE TWO). All of these 12 multiverse have the same size, shape and characteristics of our universe. Without getting too complicated for example, each one has its own Big Bang, gravity, all particles, full expansion, billions of galaxies, hundreds of billions of stars, and planets. These 12 multiverse have expanded to their full capacity and helped in the creation of our universe. So 14 billion years ago the 20 green (zone three) multiverse triangle energies and these 12 yellow (zone two) energies feeding into the center of one point created the big bang of our universe.

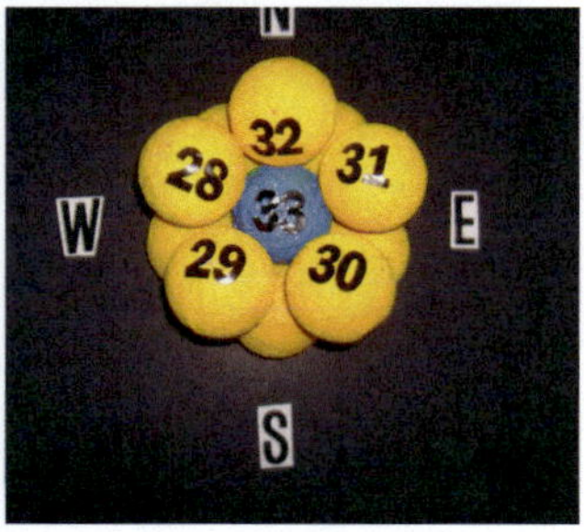

Yellow multiverse front side. Read clockwise but numerically 33 to 1. Mapping 32, 31, 30, 29, 28, 27.

(Note 27 left out of image so view of center blue universe 33 can be seen.

Front West View

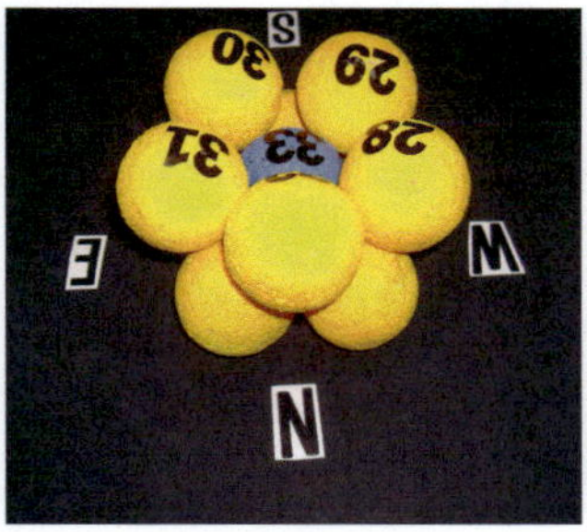

Front North View

Front East View

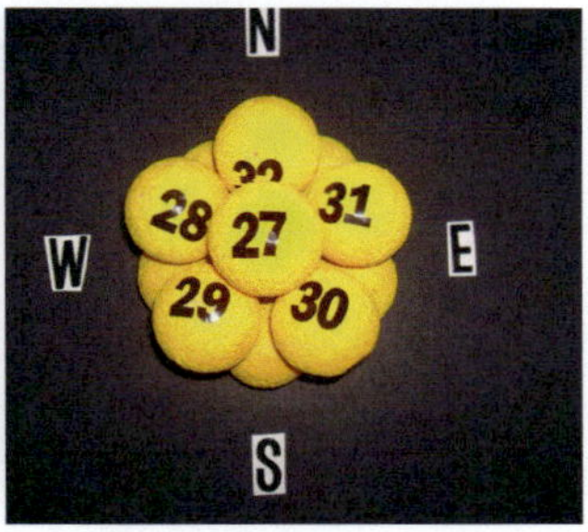

Complete Front View

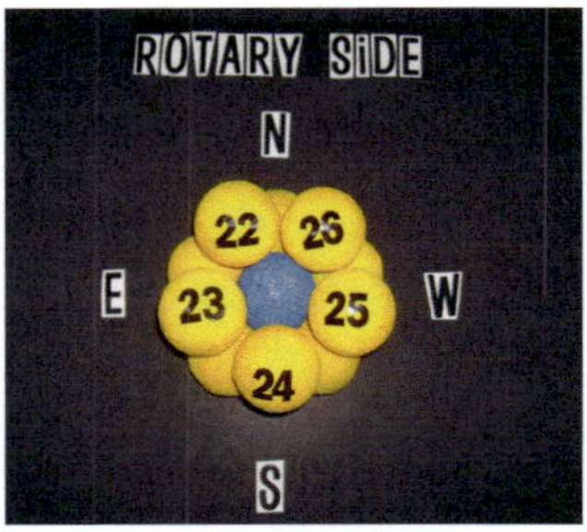

Rotary Side Mapping 26, 25, 24, 23, 22, 21 (Note: 21 purposely left out so view of rear side center Blue Universe #33 can be seen. North-West-South-East

Rotary Side East View

Rotary Side North View

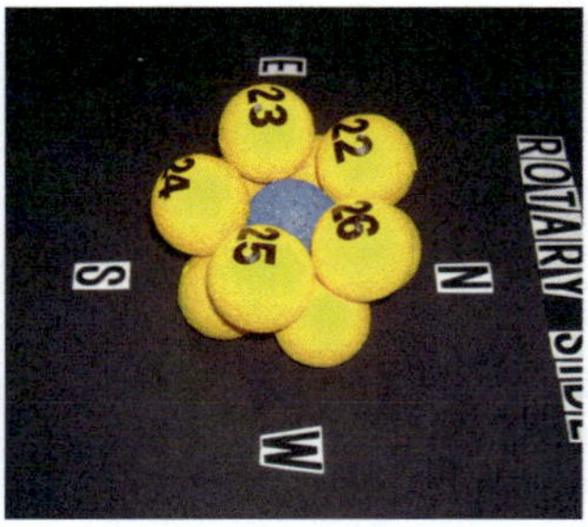

Rotary Side West View

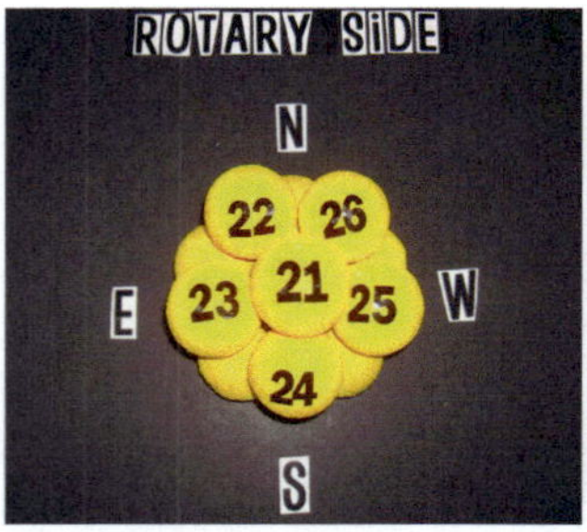

Complete Rotary View

South Close Up View

North Close Up View

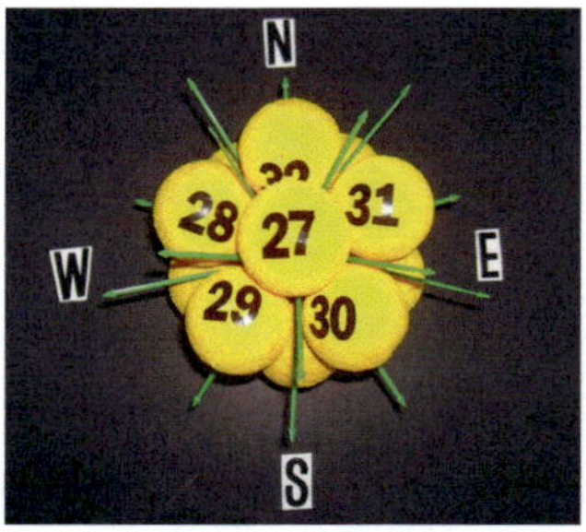

Front side view with green string energies pointed away from its point of origin

32 Multiverse energy points into one point of number 33 blue universe to create our Big Bang. Note: not all energy points are seen for viewing the center of Big Bang.

First zone blue universe
Universe number total 1

String energy to the center forms a perfect point just before energy made contact for the Big Bang

Direction of energy flow for the Big Bang

Billionth of a second after Big Bang

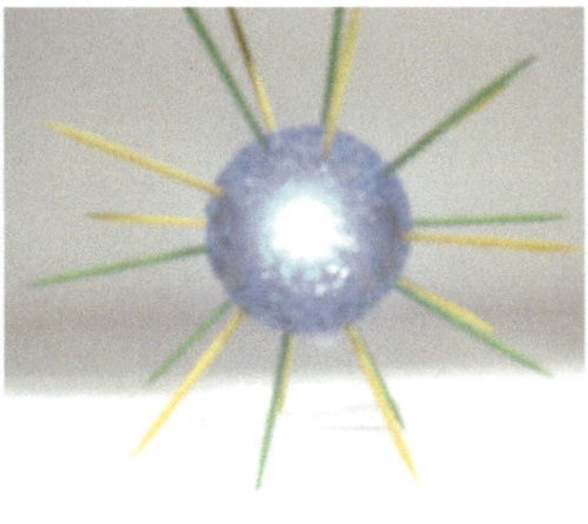

19 string energies zooming out after Big Bang

Farther back in time of the Big Bang

Zooming in view of more string energy creating Big Bang a few seconds after

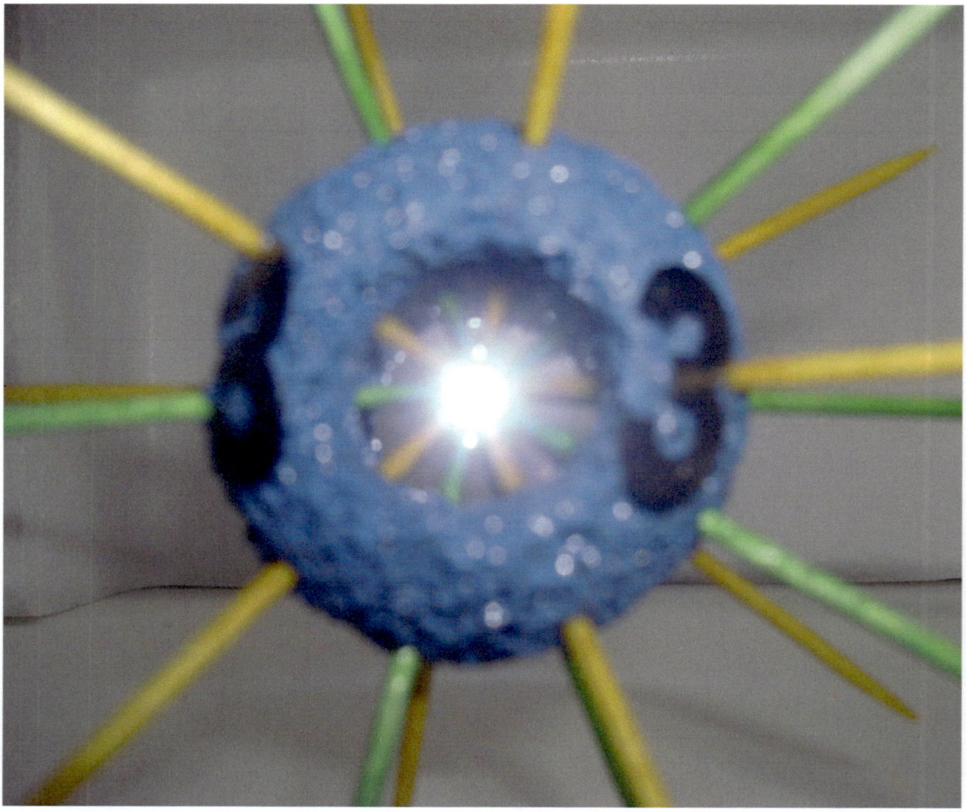

Smallest point of creation of the Big Bang. Can't see all the string energy. Front or rear only (18 string energy) unprecedented perfection.

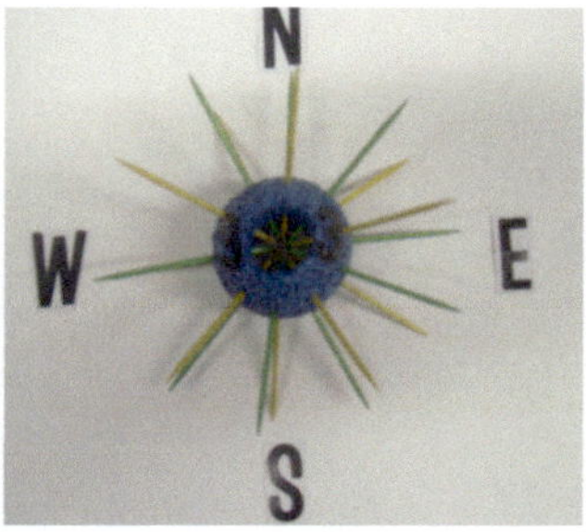

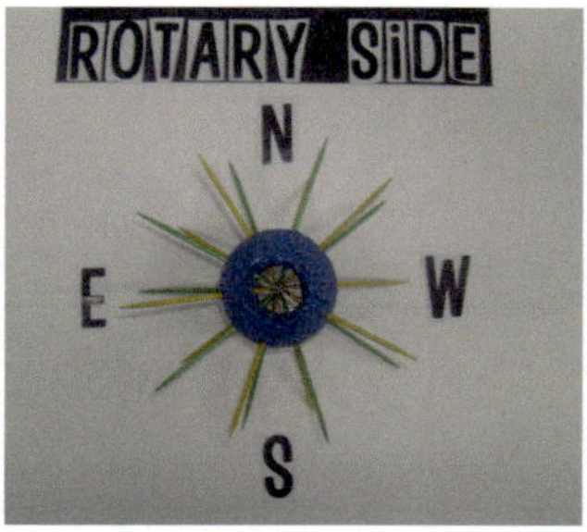

No sides. North is North no matter how you turn it.

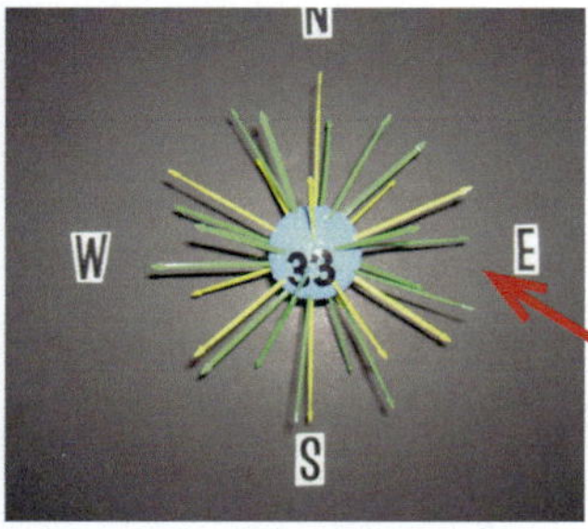

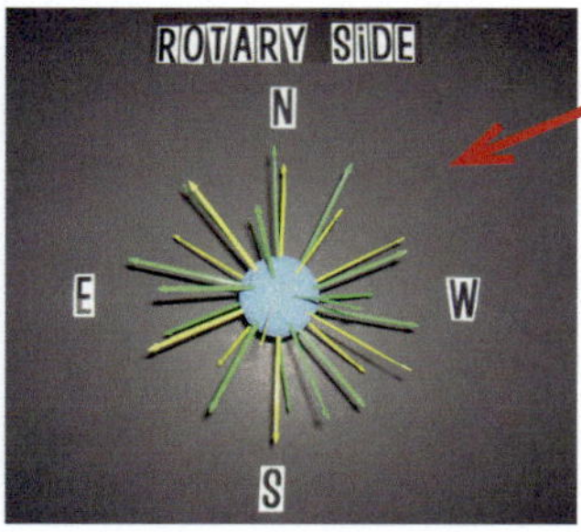

Creation Code

Rolando Rivero

$$O = O^{12}\ \Delta^{20}$$

Multiverse
12 + 20 + 1 = 33

Sugar DNA 8+8+8+8=32=1 double helix

Silicon Quartz Crystals 32 + 1 = 33

Bee chromosomes 32 +1=33

Showing all 32 points. Perfection at its best. North and South do not change.

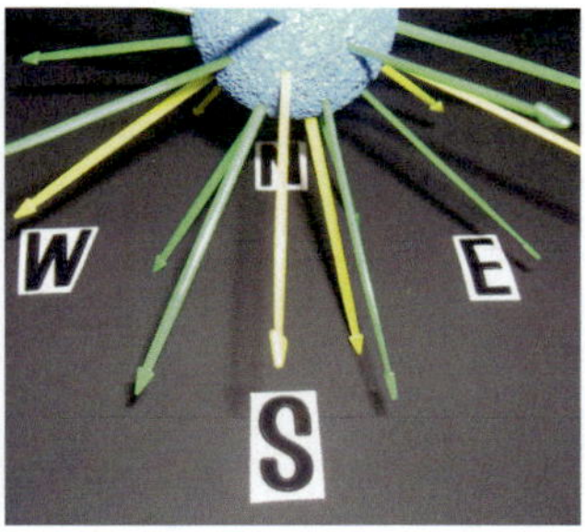

God's energy white multiverse outlined without circular shell mapped cut in half for easy viewing.

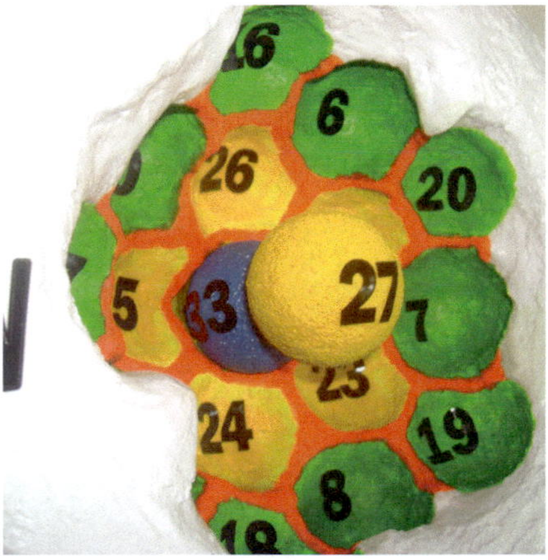

3-dimensional showing inside multiverse number 33 and 27

Big Bangs of multiverse

3-dimensional neutral area shown above. All others on this page showing dimensional 27, 31, 32 and 33.

More developed before the Big Bang

Zoomed out

Zoomed in

After Big Bang more developed multiverse

Zoomed out

Zoomed in

Map of Multiple Universe and Color Zones
First Zone Blue Universe Number Total 1
Second Zone Yellow Universe Number Total 12
Third Zone Green Universe Number Total 20
Fourth Zone Orange Universe Neutral Space Areas

Space

Archive of Multiverse Sculptures

Neutral area triangle blue 33 multiverse

Neutral area triangle multiverse encompassing blue, yellow and green

Multiverse filling neutral area

Yellow multiverse encompassing neutral area

More defined neutral area

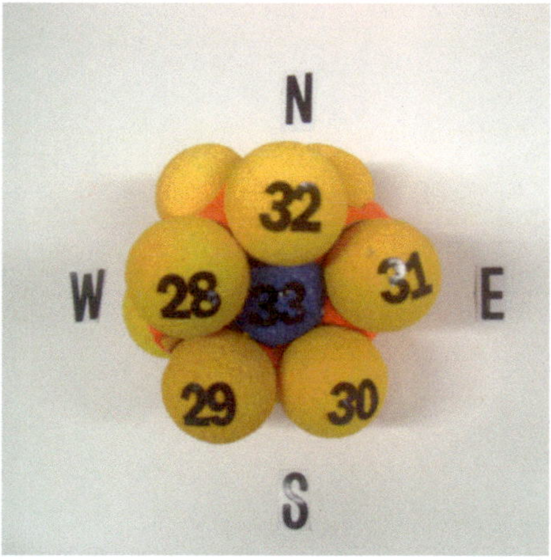

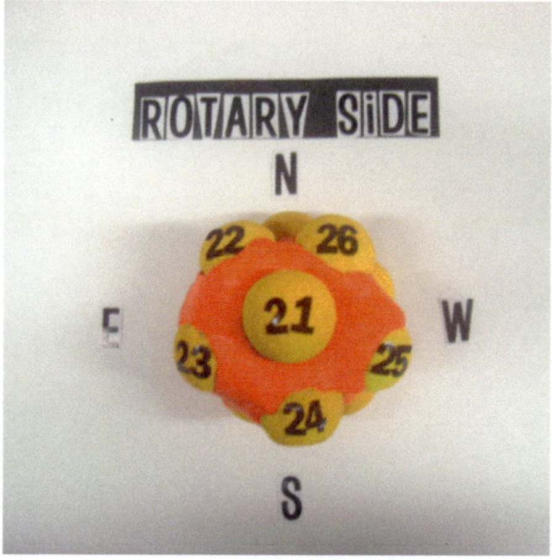

Universal pattern filling neutral area

Outer neutral area colored orange

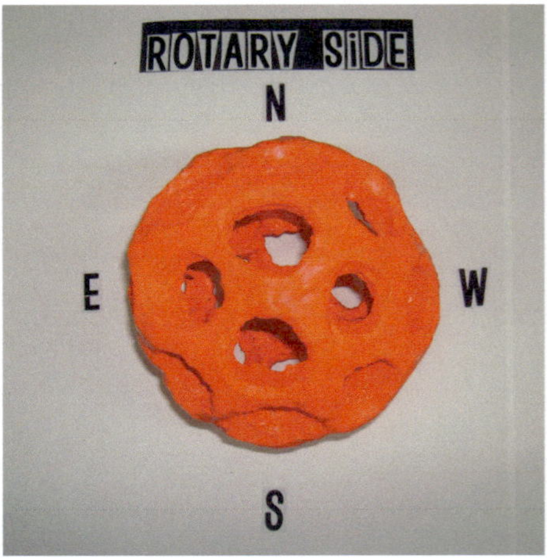

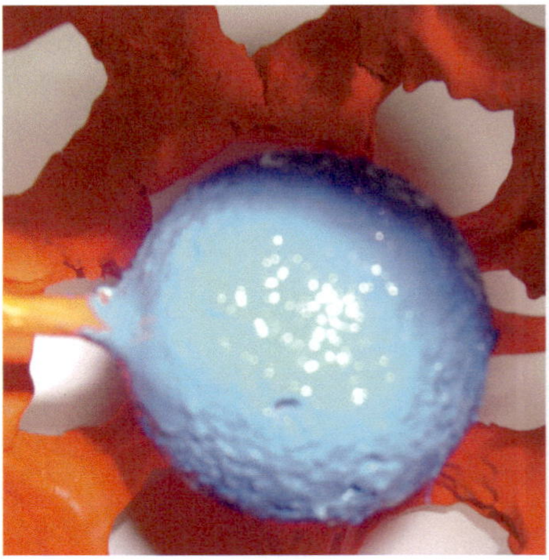

Blue universe vast space without multi-verse

The white spaces are God's space. Blue universe space without mass. Multi-verse exiting towards God's energy in white circle.

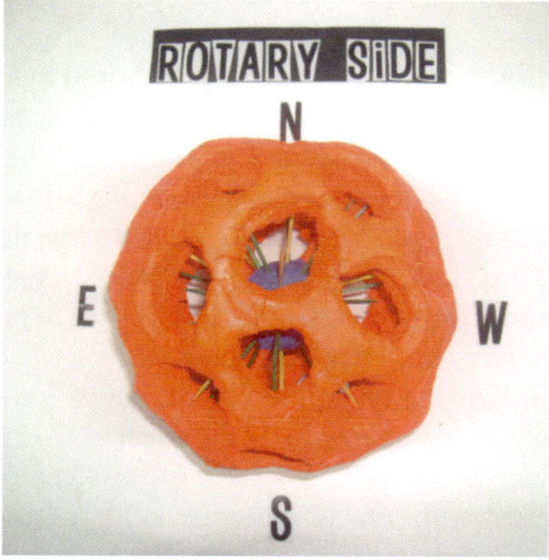

Same as at left looking at it from the outside.

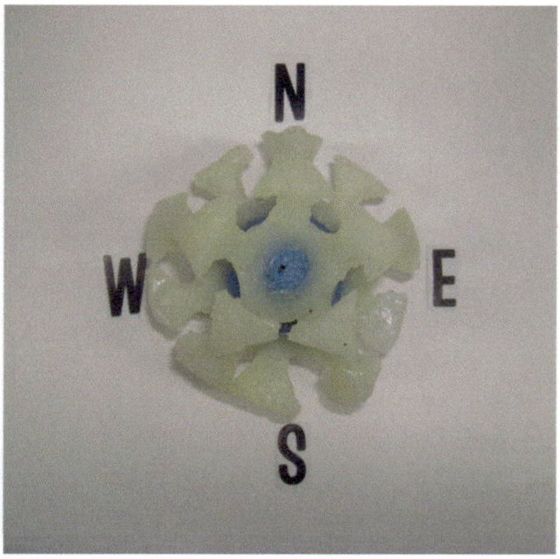

Triangle mapped neutral 20 points clear to show the direct impact at blue multi-verse.

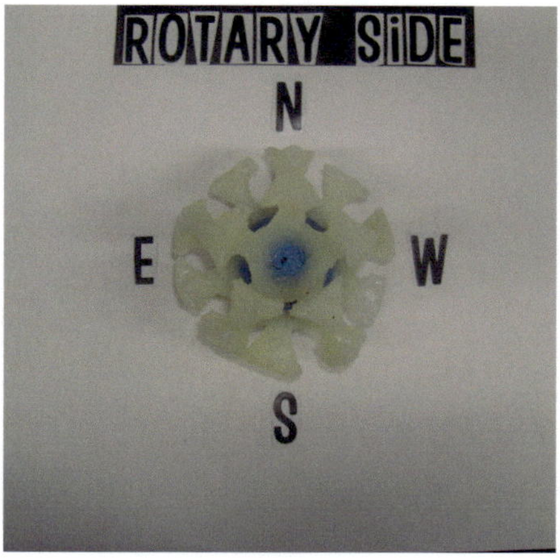

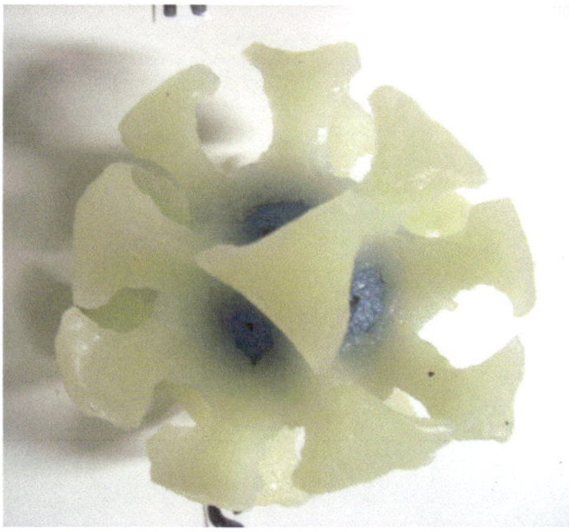

Green triangles marked 1-5.

The 32 multiverse are areas that need to be accounted for. These 32 multiverse only have certain places that touch each, there are other areas that do not touch. I call these areas neutral areas or speedy travel areas. 20 triangular neutral areas that can be used to travel from our universe to the Green zone three universes. There's a space in between the neutral areas that can be used for speedy travel, too close to the edges and you can get pulled up into one of the universes. But if you find the right spot you can use this for speedy travel. Note: in trying to identify for mapping simplicity I have colored these areas orange color. 20 triangular zone 4.

Green triangles marked 11-15.

Mapped

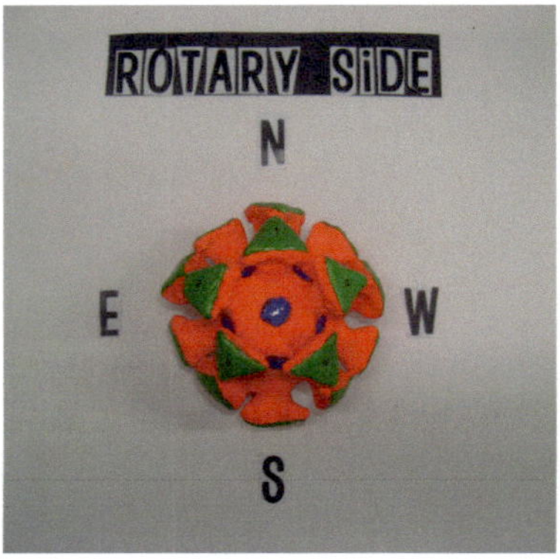

Multi-verse yellow filling #32, 29 & 30

Neutral triangle area point space multiverse

27 Covering blue multiverse

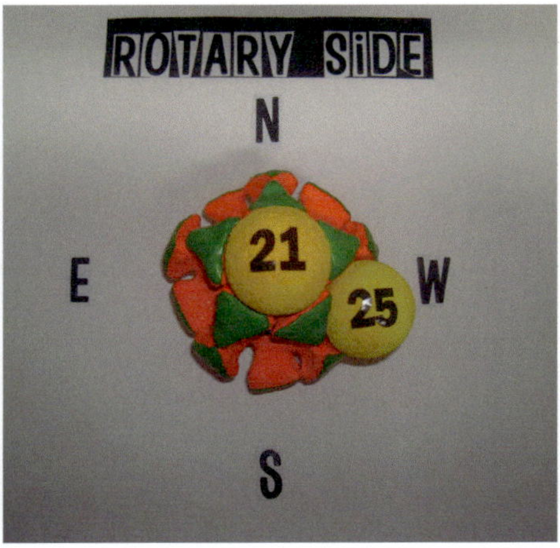

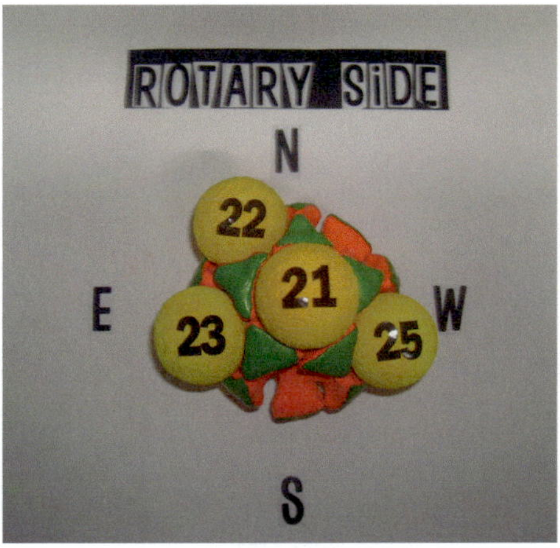

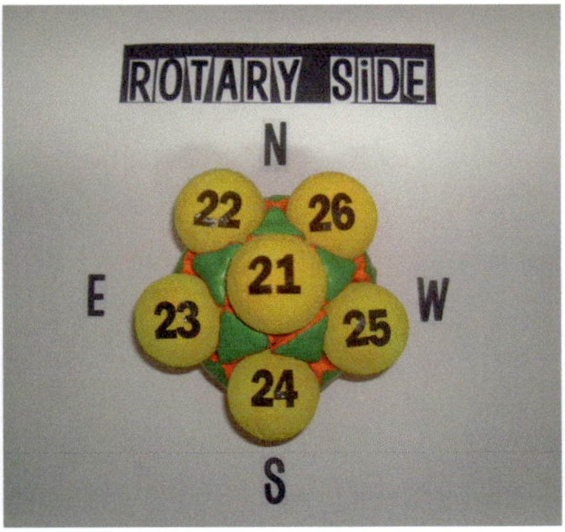

Close view from God's point of view

No matter how you move the multiverse halves they are always the same shape. Geometrical perfection.

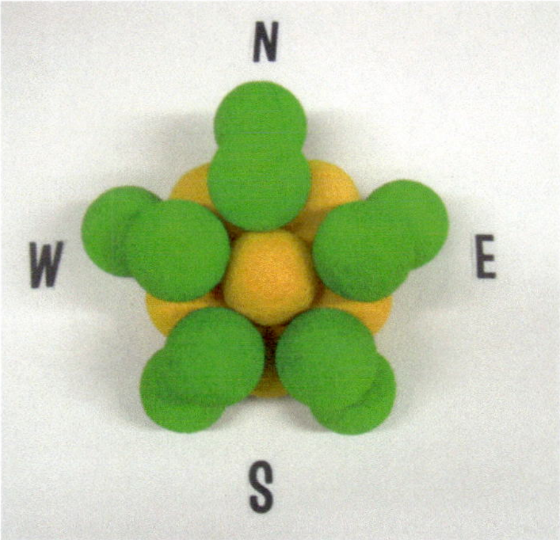

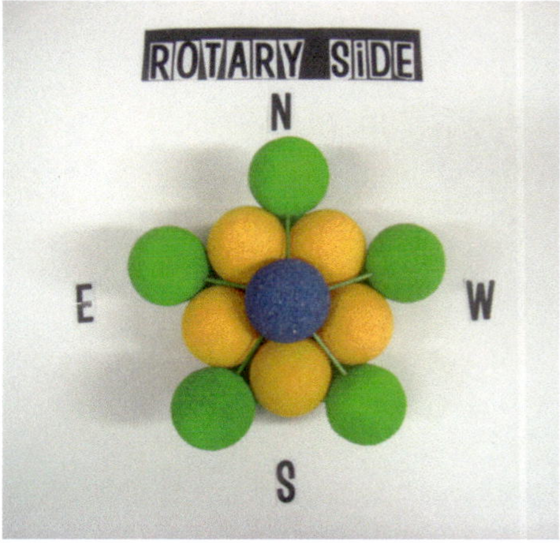

Side view

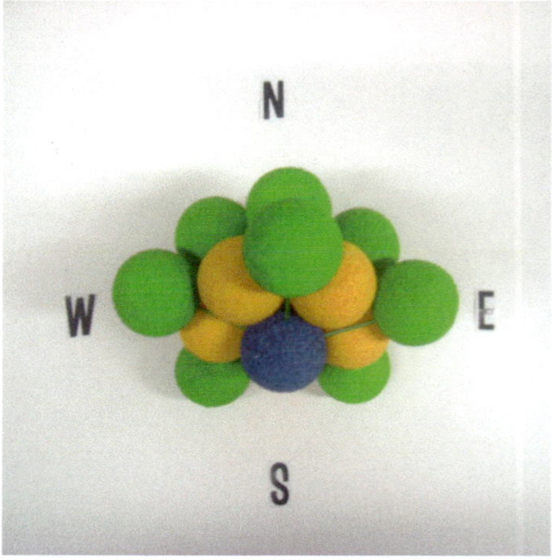

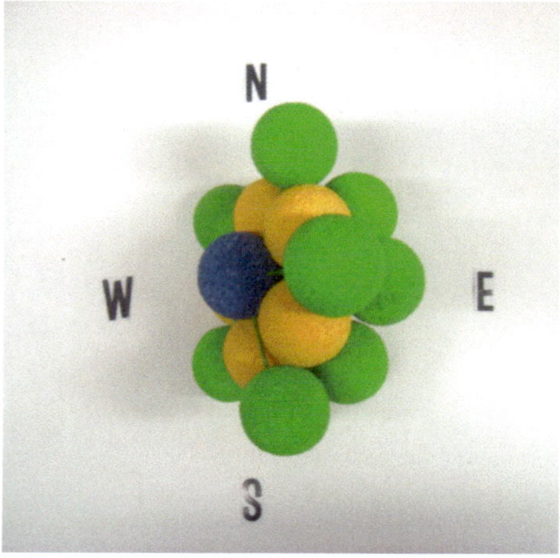

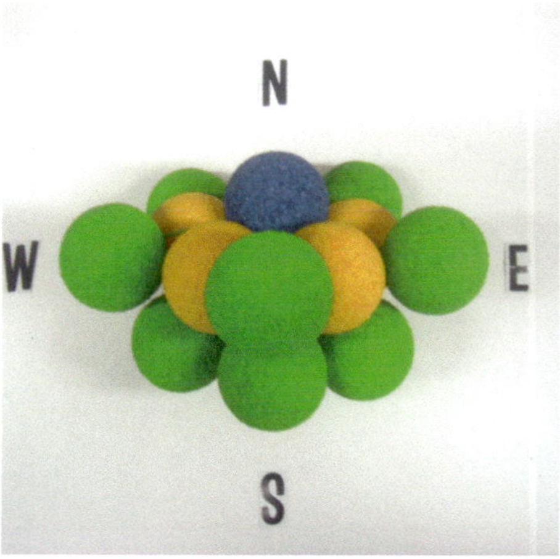

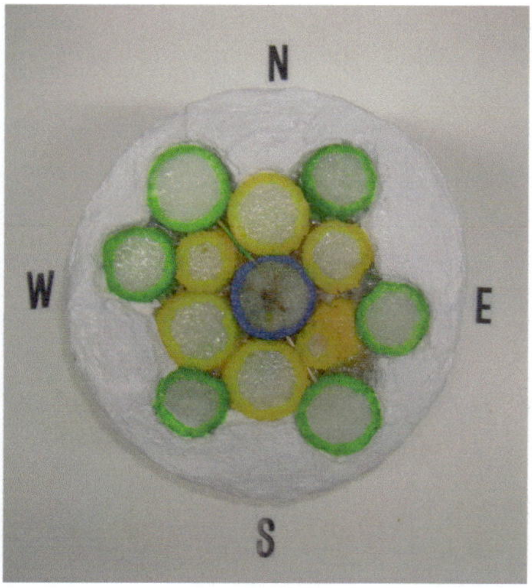

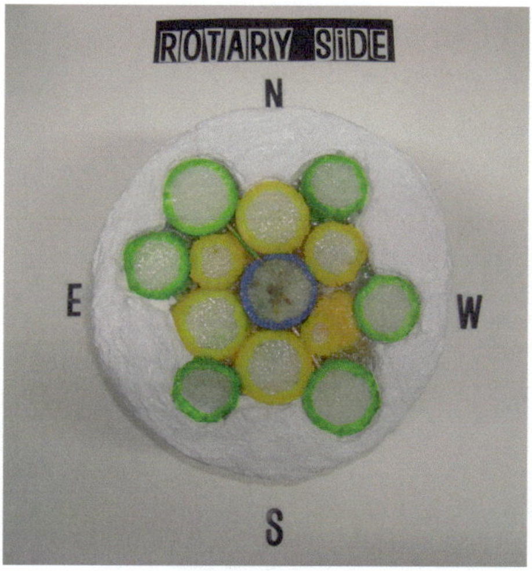

White area is God's light of the lamb, God's energy encompassing all of the multiverses mapped.

Our whole universe at its current time of expansion. It will continue to expand. Clear multi-verse to show expansion.

Universal Code

Rolando's Formula

12 + 20 = 32 +1 = 33

Rolando's 12 + 20 = 1 (Quantum Physics)

12 Faces + 20 triangle points = 1 Universe

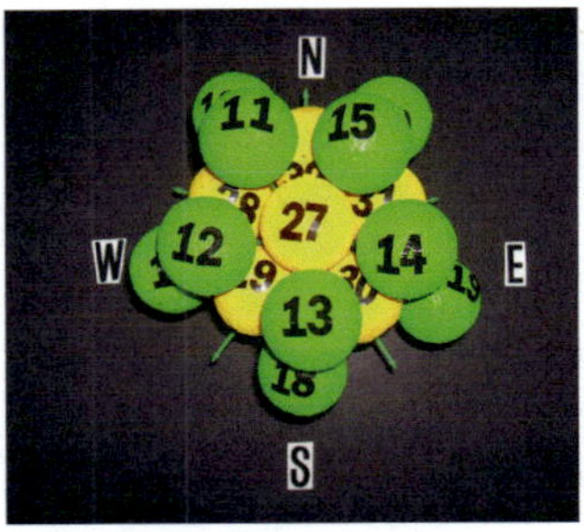

Note: trying to identify for mapping simplicity, green multiverse 1 through 20 front side, Rotary side, using North, South, East, West.

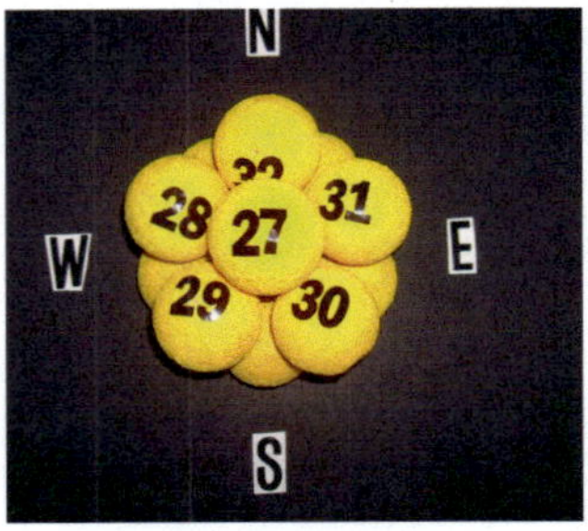

Note: trying to identify for mapping simplest, yellow multiverse 21 through 32 front side, Rotary side, using North, South, East, West

Our Universe Age and Size

According to scientific data, our universe is about 14 billion years old and our observable universe is about 156 billion across. So it is safe to say our universe has expanded less than a quarter of its container. And has about 86 billion years left. Our single number 33 blue zone 1 universe has expansion left to fill its container. The observable universe expanded about 11.14 every billions years. It is safe to say 11.14 x 100 = 114 Trillion left to fill its container.

Multiverse Age and Size

32 multiverse age and size. Since 20 green zone 3 universe were created at the same time, it is safe to say that 20 Green zone 3 multiverse is about 100 billion years old. The observable multiverse is about 114 Trillion x 20 = 2280 Quadrillion.

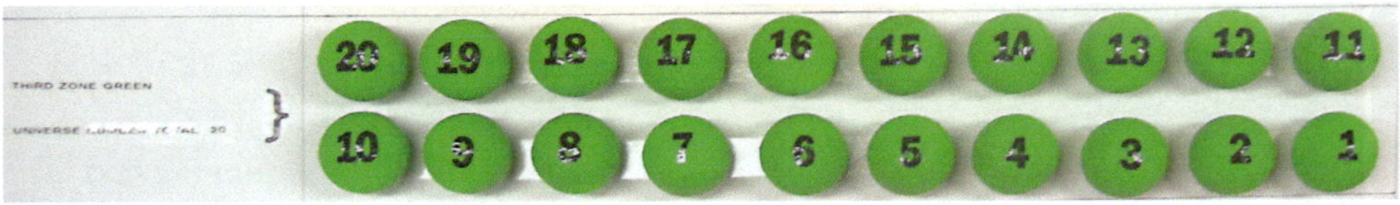

Since 12 yellow zone 2 universe were created at the same time, it is safe to say 12 yellow zone 2 multiverse is about 100 Billion years old. Observable multiverse is about 114 Trillion x 12 = 1368 Quadrillion.

Since 20 triangular neutral area zone 4 was created at the same time as the 12 yellow multiverse it is included in the 100 Billion years old. But I do have to account for its space. So it is safe to say these 20 triangular orange zone 4 neutral area is about a quarter universe space size. Observable neutral area is about 156 Billion x 20 = 3120 Trillion.

Complete Age and Size of 33 multiverse

Zone	Age	Multiverse Space size
20 green multiverse =	100 Billion Years	2280 Quadrillion
12 yellow multiverse =	100 Billion Years	1368 Quadrillion
1 blue our universe =	100 Billion Years	114 Trillion
20 orange neutral areas =		3120 Trillion
Total =	300 Billion Years	Total 3.65226 Quadrillion

Creation Code

Rolando Rivero

$$O = O^{12}\ \Delta^{20}$$

Multiverse
12 + 20 + 1 = 33

Sugar DNA 8+8+8+8=32=1 double helix

Silicon Quartz Crystals 32 + 1 = 33

Bee chromosomes 32 +1=33

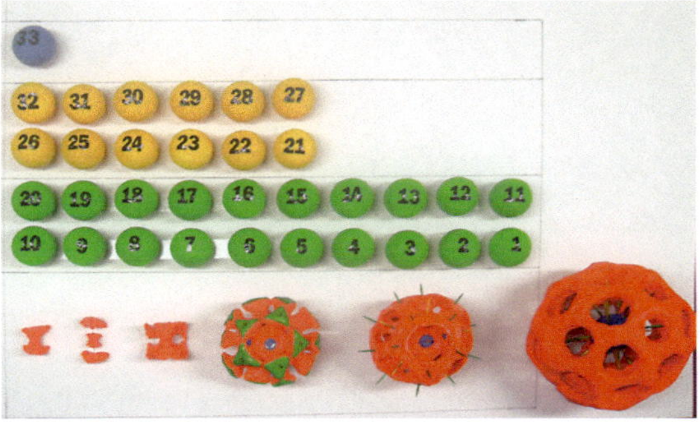

THE FIRST HOLY GRAIL CUP OF OUR MULTIVERSE

214 BILLIONS YEARS OLD THE STORY OF US

86 BILLION YEARS TO GO

Holy Grail Cup of Our Multiverse

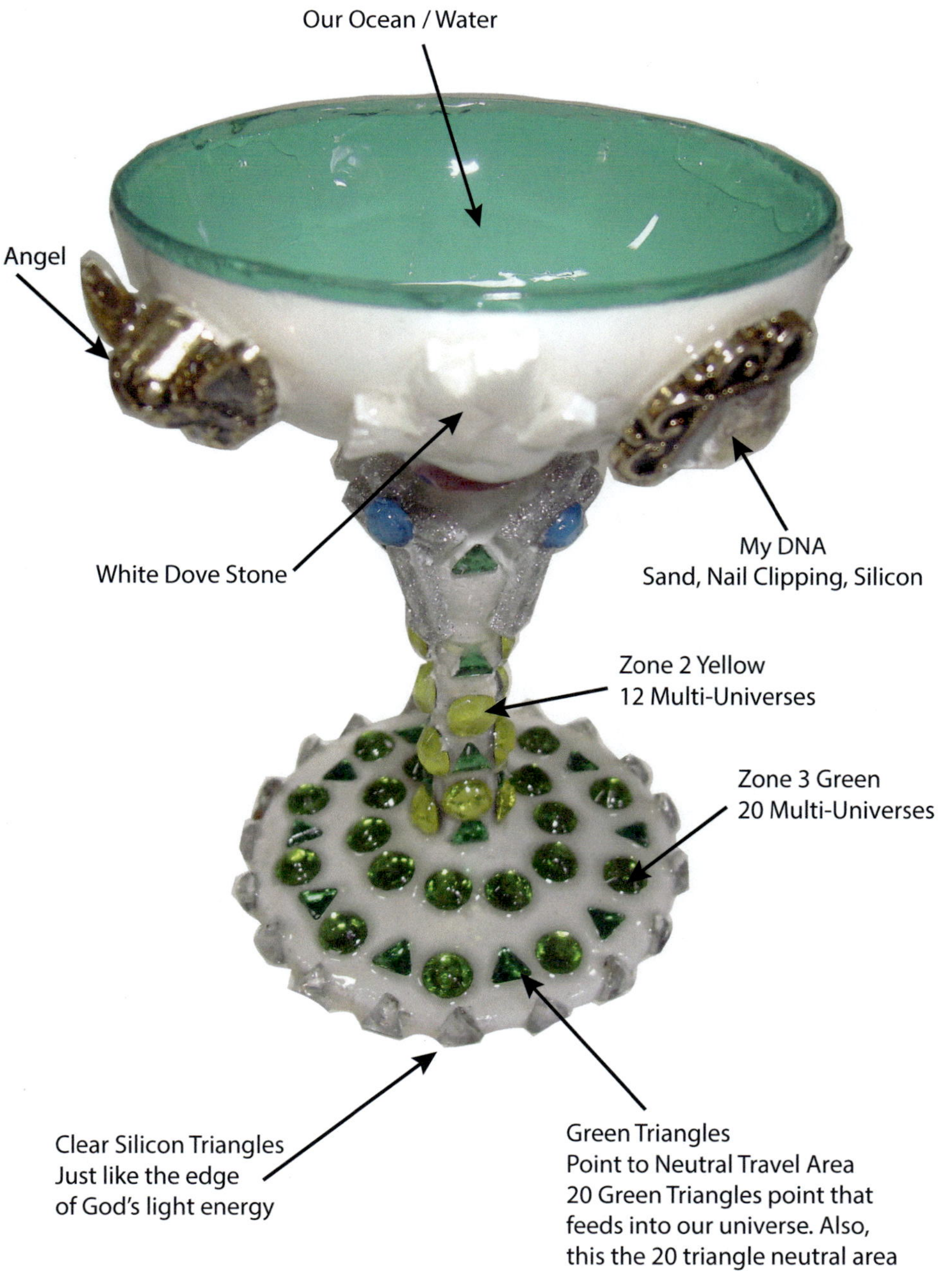

Cup embedded in portrait on page 64

Our Present Universe

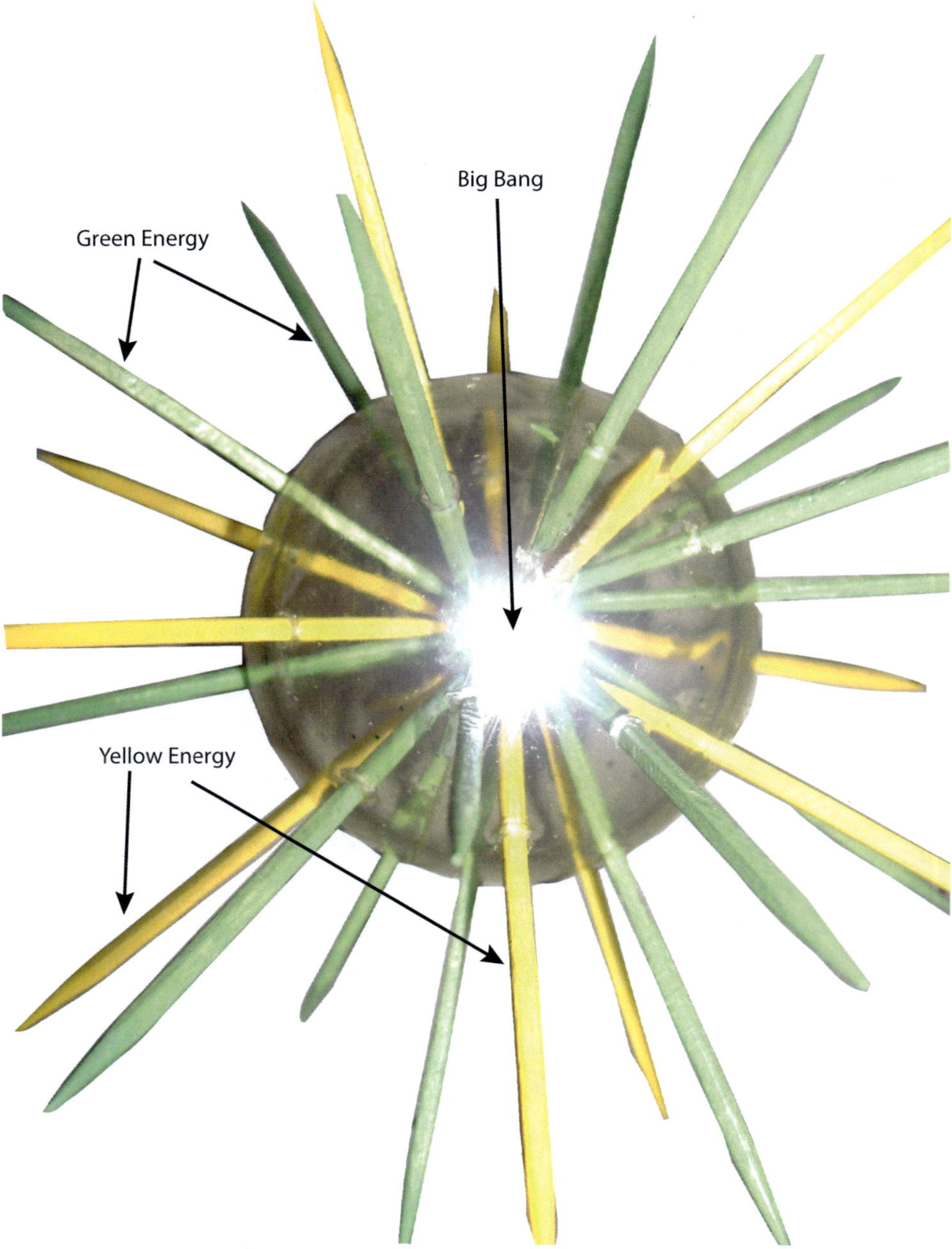

Our whole universe at its current time, 32 multiverse energy points fanning our universe's continuing expansion.

All multiverses pulsating, vibrating, string energies to one point to create our big bang expansion. Fanning our universe in constant expanding motion.

Because of the energies colliding on the edges, a curtain of elements was formed. These elements created cosmic rays of light and heat for a long time. Secondly, because of this certain areas of the outer edges were changing. Some areas were soft while some were hard. A gassy, milky, silicon crystal. And the patterns were consistent. Time would pass and now in a lot of areas of triangular shapes with curved sides, tetrahedrons were formed. And in these tetrahedrons, elements were constantly pushing and pulling hand in hand tug-o-war battle. Dark pressure void with quantum dark energy, gravity giving a helping hand. Because these tetrhedrons have 3 curved sides, these energies were bouncing off in a circular rotation creating circular craters in motion. And between the void's pressure, the first 20 green multiverses were born and at that time 12 yellow multiverses were taking shape. But these 12 yellow multiverses had smaller craters on edge and more help from the green multiverses. So between small void edge craters, green multiverses and pressurized void the inner sphere formed 12 yellow multiverse. And the full expansion of 20 green multiverse and the full expansion of 12 yellow multiverse – 32 pulsating string energies all together entered into the center of the one pressurized dark void. Each of the multiverses would have their own four forces – Weak Nuclear, Electromagnetism, Gravity and Strong Nuclear plus similar to our universe their own big bang, inflation and expansion. In the center point of this dark void, all 32 pulsating string energies collided to become our Big Bang creation. The beginning of our universe. Today these energies are still feeding our universe, keeping it fanning in motion and expansion.

The Uniformity Origin of Life

Today many scientists still hide their belief in a higher power, keeping up the old thought that there is no room for God in the scientific realm. Yet there are those who believe that faith can provide the basis for scientific discovery. A University of Technology in Tehran physicist, Mehdi Golshani, believes that natural phenomena are "God's signs in the universe," and that studying them is nearly a religious obligation. He believes that the Koran asks humans to travel the earth, then visualize how God initiated the creation. He also believes that research is an act of worship, because it reveals so much information surrounding the wonders of God's creation.

That same thought process can be seen in Judaism. According to Carl Feit, Maimonides said that the only pathway to achieve a love of God is by understanding the works of his hand, which is the natural universe. Learning how the universe functions is critical to a person of faith, after all, this is the world God created.

Rather than look at religion and science as adversaries we should view science and spirituality as having a common quest, which is a search for truth. For example, both religion and science show us that the universe obeys certain rules—laws to which all things must adhere. These laws are precise, and many of them are mathematical in nature. Everything in the universe, every plant and animal, every rock, every particle of matter or light wave, is bound by laws which it has no choice but to obey. The Bible tells us that there are laws of nature—"ordinances of heaven and earth" (Jeremiah 33:25).

These natural laws are The Law of Life (Biogenesis), The Laws of Chemistry, The Laws of Planetary Motion, The Laws of Physics, The Laws of Mathematics, The Laws of Logic, and The Uniformity of Nature.

The Law of Life (biogenesis): Life always comes from life. This is what observational science tells us: organisms reproduce other organisms after their own kind.

The Laws of Chemistry: The properties of elements and compounds are not arbitrary. The elements are logically organized into a periodic table based on their physical properties. This is not happenstance. Atoms and molecules have their various properties because their electrons are bound by the laws of quantum physics. In other words, chemistry is based on physics. If the laws of quantum physics were just a bit different, atoms might not even be possible.

Learning how the universe functions is critical to a person of faith, after all, this is the world God created.

The Laws of Planetary Motion: Creation scientist Johannes Kepler discovered that the planets in our solar system obey three laws of nature. He found that planets orbit in ellipses (not perfect circles as had been previously thought) with the sun at one focus of the ellipse; making any given planet sometimes closer to the sun than at other times. Another creation scientist (Sir Isaac Newton) discovered that Kepler's laws could be derived mathematically from certain laws of physics—specifically, the laws of gravity and motion (which Newton himself formulated).

The Laws of Physics: Describes the behavior of the universe at its most fundamental level. There are many different laws of physics. They describe the way the universe operates today. Some laws of physics describe how light propagates, how energy is transported, how gravity operates, how mass moves through space, and many other phenomena. The laws of physics are normally mathematical in nature; some can be described with a concise formula, such as Einstein's E=mc2. The simple formula F=ma shows how an object with mass (m) will accelerate (a) when a net force (F) is applied to it. Every object in the universe consistently obeys these rules. The laws of physics (along with their associated constants) are fine-tuned in just the right way so that life, particularly human life, is possible. This fact is called the "anthropic principle."[5]

The Laws of Mathematics: Addition, the transitive property, the commutative properties of addition and multiplication, the binomial theorem, and many others are laws of mathematics. Like the laws of physics, there are some laws and properties of mathematics that can be derived from other mathematical principles. But unlike the laws of physics, the laws of mathematics are abstract; they are not "attached" to any particular part of the universe. One can imagine a universe where the laws of physics are different, but it is difficult to imagine a (consistent) universe where the laws of mathematics are different.

The Laws of Logic: All the laws of nature, from physics and chemistry to the law of biogenesis, depend on the laws of logic. Atheists cannot account for the laws of logic, even though they must accept that they exist in order to do any rational thinking. According to the Bible, God is logical. For example, the law of non-contradiction reflects God's nature; God cannot lie (Numbers 23:19) or be tempted with evil (James 1:13) since these things contradict His perfect nature. Because we are made in God's image, we instinctively know the laws of logic.

The Uniformity of Nature: Laws of nature are uniform. They do not change arbitrarily, and they are applicable throughout the whole cosmos. The laws of nature apply to the future just as they have applied during the past; this is one of the most basic assumptions in all of science. Why can we depend on the laws of nature to apply consistently throughout the course of time? The Bible provides the answer. God never changes, and so He upholds the universe in a consistent, uniform way throughout time (Jeremiah 33:25).

So if we delve into the law of physics and how it applies to making human life possible, we see that the Earth must have a magnetic field to protect its atmosphere from dehydration. We also know that water is absolutely necessary for life to exist. Some scientists believe that Mars' desert-like environment may have been caused due to its magnetic field somehow dying. So did we climb out of the primordial soup as some evolutionists would have us believe? Or was there a higher power at play?

Let's take a closer look at the magnetic field. Recent scientific analysis of tiny iron minerals found inside 3.45 billion year old dacite rocks from the Barberton mountains in

[5] Anthropic comes from the Greek word for man, anthropos.

South Africa have provided evidence of a weaker field than what we have today. Findings of a study by led Professor John Tarduno of the University of Rochester suggest that the Earth contained much more water than the planet does today. Which brings me to question whether we are in danger of losing our own planet's magnetic field of protection. Further solidifying my hopes that the world of science will begin to look for solutions either to preserve that magnetic field or to begin the process of building the systems and pathways to other universes that would be able to sustain us.[6]

While a "standard model" of the origin of life has not yet been determined, most currently accepted models draw at least some elements from the framework laid out by the Oparin-Haldane hypothesis. Both Aleksandr Oparin, and John Haldane, independently suggested that if Earth's primitive atmosphere was reducing rather than oxygen-rich, and an appropriate supply of energy, such as lightning or ultraviolet light, was provided then a wide range of organic compounds could be synthesized.

Oparin suggested that organic compounds may have undergone a series of reactions leading to molecules becoming more and more complex. He proposed that the molecules were able to form colloid aggregates, or 'coacervates', in an aqueous environment. The coacervates were then able to absorb and assimilate organic compounds from their environment in a method similar to metabolism. And that they would have taken part in evolutionary processes that would eventually lead to the first lifeforms.

Haldane's theories about the origin of life were quite similar to Oparin's. Haldane suggested that the primordial sea provided a vast chemical laboratory which was powered by the Sun's energy. Earth's atmosphere was oxygen free, and that the combination of carbon dioxide, ammonia and ultraviolet radiation created a host of organic compounds. The sea then became a 'hot dilute soup' containing large populations of organic monomers and polymers. Haldane imagined that groups of monomers and polymers acquired lipid membranes, and that further developments eventually led to the first living cells.

Haldane coined the term 'prebiotic soup', and this became a powerful symbol of the Oparin-Haldane view of the origin of life.

Biochemist Robert Shapiro has summarized the "primordial soup" theory of Oparin and Haldane in its "mature form" as follows:[7]

The early Earth had a chemically reducing atmosphere.

This atmosphere, exposed to energy in various forms, produced simple organic compounds ("monomers"). These compounds accumulated in a "soup", which may have been concentrated at various locations (shorelines, oceanic vents etc.). By further transformation, more complex organic polymers – and ultimately life – developed in the

[6] http://www.wired.com/2010/03/earths-magnetic-field-is-35-billion-years-old/

[7] Shapiro, Robert (1987). Origins: A Skeptic's Guide to the Creation of Life on Earth. Bantam Books. p. 110. ISBN 0-671-45939-2.

Notes from http://en.wikipedia.org/wiki/Origin_of_life

http://news.harvard.edu/gazette/story/2008/10/nyu-chemist-robert-shapiro-decries-rna-first soup. While steps 1-3 have been basically observed experimentally, step 4 has been criticized as simplistic - a stage of "then magic happens".

Shapiro suggests that a metabolic reaction, rather than DNA or RNA, could be the first process in the start of life. This would mean life in the universe has the potential to be quite common since an extraordinary event would not be necessary to create it. Shapiro further implies that this metabolic reaction would need 5 items to create life, including a boundary to contain the ingredients, perhaps a rock-bound compartment such as a geode. In addition, a metabolism-like reaction would require an energy source; a chemical group capable of allowing adaptation and evolution, a coupling of the energy to a "driver reaction"; and finally reproduction. Shapiro believed this was a normal consequence of the laws of nature and that this process was actually common throughout the universe.

Clay Hypothesis

A model for the origin of life based on clay was forwarded by A. Graham Cairns-Smith of the University of Glasgow in 1985 and explored as a plausible illustration by several scientists. The Clay hypothesis suggests that complex organic molecules arose gradually on a pre-existing, non-organic replication platform of silicate crystals in solution. According to the Bible, humankind was created from the dust of the earth and from the breath of God. Genesis 2:7 says *"And the Lord God formed man of the dust of the ground, and breathed into his nostrils the breath of life; and man became a living soul."* Could this clay hypothesis be the scientific proof that we were made from the Earth by God as the Bible says?

Cairns-Smith is a trenchant critic of other models of chemical evolution. However, he admits that like many models of the origin of life, his own also has its shortcomings.

In 2007, Kahr and colleagues reported their experiments that tested the idea that crystals can act as a source of transferable information, using crystals of potassium hydrogen phthalate. "Mother" crystals with imperfections were cleaved and used as seeds to grow "daughter" crystals from solution. They then examined the distribution of imperfections in the new crystals and found that the imperfections in the mother crystals were reproduced in the daughters, but the daughter crystals also had many additional imperfections. For gene-like behavior to be observed, the quantity of inheritance of these imperfections should have exceeded that of the mutations in the successive generations, but it did not. Thus Kahr concluded that the crystals, "were not faithful enough to store and transfer information from one generation to the next".[8][9]

Quartz is the second most abundant mineral in the Earth's continental crust, after feldspar. It is made up of a continuous

[8] Bullard T, Freudenthal J, Avagyan S, Kahr B (2007). "Test of Cairns-Smith's crystals-as-genes hypothesis". Faraday Discuss. 136: 231–45. Bibcode:2007FaDi..136..231B. doi:10.1039/b616612c. Caroline Moore (16 July 2007). "Crystals as genes?". Chemical Science.

[9] Wikipedia.org/wiki/silicon_life

framework of SiO_4 silicon–oxygen tetrahedra, with each oxygen being shared between two tetrahedra, giving an overall formula SiO_2.

British biochemist Alexander Graham Cairns-Smith believed life evolved from simple inorganic materials, particularly crystals, rather than from organic molecules.[10] His theory was simple and dealt with two familiar properties of crystals that are much like what we consider essential properties for labeling anything "living"- reproduction and natural selection. Crystals are periodic, extending regularly in infinite planes, "reproducing" in three dimensions. Crystals are also subject to defects and impurities. The beauty of crystallization is that impurities or defects always exist. Cairns-Smith recognized that the impurities or defects may supply some benefit like better 'stickiness' or other mechanical properties, and would therefore be propagated in the way that beneficial mutations are propagated through evolution. Slowly, the old, decrepit crystal will be left behind and a new impure crystal with its superior properties would take over. Ergo crystal evolution.

According to Cairns-Smith theory of life from clay, silicon dioxide in all its various forms is the key. For billions of years on earth silicon dioxide has been extremely abundant. Secondly, it comes in an amazing variety of polymorphs, allotropes and geometries. Silicon can mix and match with numerous cations and anions and form quixotic 3D structures. It can act as a scaffold for many other substances.

Cairns-Smith believed that crystals of silicon easily harbor organic impurities in them. Over time, these impurities grow along with the crystal. At some point, the organic molecules would have an evolutionary benefit possibly due to their greater flexibility, branching power, and stickiness caused by hydrophobic interactions and multiple bonding characteristics. Eventually the organic molecules would simply grow faster and stronger and leave their old siliconian nature. Eventually it would become an organic crystal. Now think about DNA, RNA, proteins forming from such organic entities.

Scientists have come to believe that life began in hydrothermal vents in the deep sea, but a 2012 study led by Armen Mulkidjanian of Germany's University of Osnabrück, suggests that inland pools of condensed and cooled geothermal vapor have the ideal characteristics for the origin of life. In 2002, some scientists discovered that by adding a montmorillonite clay to a solution of fatty acid micelles (lipid spheres), the clay sped up the rate of vesicles formation 100-fold.[10] So this one mineral can get precursors (nucleotides) to spontaneously assemble into RNA and membrane precursors to assemble into membrane. Oceanic vent systems and other hydrothermal systems have a zonal structure reflected in ancient volcanogenic massive sulfide deposits (VMS) of hydrothermal origin. They reach many kilometers in diameter and date back to the Archean eon. Most abundant are pyrite (FeS_2), chalcopyrite ($CuFeS_2$), and sphalerite (ZnS), with additions of galena (PbS) and alabandite (MnS). ZnS and MnS have a unique ability to store radiation energy, e.g. provided by UV light.

Factors supportive of an important role for RNA in early life include its ability to act both to store information and to catalyze chemical reactions (as a ribozyme); its many important roles as an intermediate in the expression and maintenance of the genetic information (in the form of DNA) in modern organisms; and the ease of chemical synthesis

[10] http://wavefunction.fieldofscience.com/2008/04/its-clay-time-origins-of-silicon-based.html

of at least the components of the molecule under the conditions that approximated the early Earth. Relatively short RNA molecules have been artificially produced in labs, which are capable of replication.[12]

Comparison of DNA, RNA and proteins[11]

	DNA	RNA	Proteins
Encodes genetic information	Yes	Yes	No
Catalyzes biological reactions	No	Yes	Yes
Building blocks (type)	Nucleotides	Nucleotides	Amino Acids
Building blocks (number)	4	4	20
Strandedness	Double	Single	Single
Structure	Double Helix	Highly Complex	Highly Complex
Stability to degradation	Extremely High	Variable	Variable
Repair systems	Yes	No	No

All living organisms are dependent on three types of very large molecules for essentially all of their biological functions. These molecules are DNA, RNA and proteins, and are classified as biological macromolecules. Without DNA, RNA and proteins, all known forms of life could not exist. This is because each molecule plays an indispensable role in biology. DNA is an informational macromolecule that encodes the complete set of instructions (the genome) that are required to assemble, maintain, and reproduce every living organism. Proteins are responsible for catalyzing the myriad biochemical reactions that are required to provide food and energy for every organism, and to carry out all of the additional functions of any given organism, for example movement, neural function, vision, or photosynthesis. RNA is multi-functional, but its primary responsibility is to make proteins, according to the instructions encoded within a cell's DNA.[13] The simple summary is that DNA makes RNA, and then RNA makes proteins. Deoxyribonucleic acid, otherwise known to us as DNA is a molecule which is responsible for creating the genetic instructions used in the development and functioning of all known living organisms as well as many viruses. The majority of DNA molecules are double-stranded helices, made up of two long biopolymers consisting of simpler units called nucleotides—each one of these is composed of a nucleobase (guanine, adenine, thymine, and cytosine), as well as a backbone made of alternating sugars (deoxyribose) and phosphate groups (related to phosphoric acid), with the nucleobases (guanine, adenine, thymine, and cytosine) attached to the sugars.

The chemical structure of RNA is very similar to that of DNA, but differs in three main ways. First, RNA is a single-stranded molecule in many of its biological roles, while DNA is a double stranded molecule. RNA also has a much shorter chain of nucleotides. DNA contains deoxyribose, while RNA contains ribose (there is no hydroxyl group attached to the pentose ring in the 2' position in deoxyribose). RNA is less stable than DNA because it is more prone to hydrolysis due to these hydroxyl groups.

DNA was first identified and isolated by Friedrich Miescher and the double helix structure of DNA was first discovered by James Watson and Francis Crick. DNA nanotechnology uses the unique molecular recognition properties of DNA and other nucleic acids to create self-assembling branched DNA complexes with useful properties. DNA is thus used as a structural material rather than as a carrier of biological information. This has led to the creation of two-dimensional periodic lattices (both tile-based as well as using the "DNA origami" method) as well as three-dimensional structures in the shapes of polyhedra.

[11] http://en.wikipedia.org/wiki/DNA,_RNA_and_proteins:_The_three_essential_macromolecules_of_life

[12] http://en.wikipedia.org/wiki/Abiogenesis

[13] http://en.wikipedia.org/wiki/DNA

DNA Creation Code

Nanomechanical devices and algorithmic self-assembly have also been demonstrated, and these DNA structures have been used to template the arrangement of other molecules such as gold nanoparticles and streptavidin proteins.

DNA (deoxyribonucleic acid) is a chemical consisting of a sequence of hundreds of millions of nucleotides found in the nucleus of cells; DNA is shaped like a double stranded helix. It is the genetic blueprint from which RNA and protein are made. All genetic information about an individual is contained within its DNA. DNA nucleotide consists of a phosphate group, a sugar and a nitrogenous base. The phosphate group and sugar are the same in every nucleotide, but there are four different nitrogenous bases: guanine, adenine, thymine and cytosine. They are often abbreviated by the first letter of each nitrogenous base: G, A, T and C. 'Reading' the DNA code tells a cell how to make proteins that it can use to perform various functions necessary for life. The single-ring nitrogenous bases, thymine and cytosine, are called pyrimidines, and the double-ring bases, adenine and guanine, are called purines. You can remember this easily by using the mnemonic "All Gods are pure", meaning Adenine and Guanine are Purines. That leaves the other two as pyrimidines.

Creation Code

Rolando Rivero

$$O = O^{12}\ \Delta^{20}$$

Multiverse
12 + 20 + 1 = 33

Sugar DNA 8+8+8+8=32=1 double helix

Silicon Quartz Crystals 32 + 1 = 33

Bee chromosomes 32 +1=33

The chemistry of the nitrogenous bases is really the key to the function of DNA. It allows something called complementary base pairing. Cytosine can form three hydrogen bonds with guanine, and adenine can form two hydrogen bonds with thymine. Or, put more simply, C bonds with G and A bonds with T. It's called complementary base pairing because each base can only bond with a specific base partner. The structures complement each other much like a lock and key. Cytosine will only bond with Guanine and Adenine will only bond with Thymine in DNA. This complementary base pairing, the hydrogen-bonded nitrogenous bases is often referred to as base pairs.

Due to the alternating nature of the phosphate groups and sugars in the backbone of nucleic acids, a nucleic acid strand has directionality. The end of a nucleic acid where the phosphate group is located is called the 5' end. The end of the nucleic acid where the sugar is located is called the 3' end. Finally, DNA strands are anti-parallel, meaning that the strands in a DNA molecule are parallel, but are oriented in opposite directions. Essentially, the 5' end of one strand pairs with the 3' end of the other strand.

Several research groups have shown that DNA itself can be used to store or manipulate data, and the juggling of DNA in a test tube or within bacteria has been shown to solve simple computational tasks. Researchers at IBM have discovered how to combine lithographic patterning and DNA self-assembly to create semiconductors that build themselves into chips that are smaller, more efficient and less expensive than conventional methods.

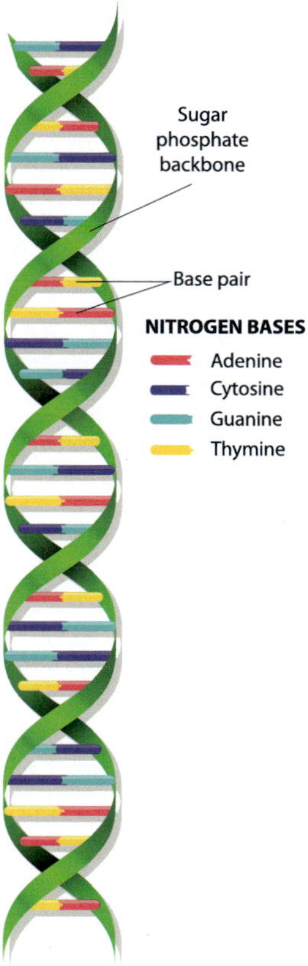

Is it possible that there exist undiscovered life-forms that use something other than carbon for their structures? Something other than water as a solvent? Something other than RNA and DNA? Such noted scientists as geneticist J.B.S Haldane, microbiologist Peter Sneath, biochemist and science fiction writer Isaac Asimov and astronomer Carl Sagan are among the many who have considered that possibility. While Sagan felt carbon appears more abundant and versatile as a chemical, he considered germanium and silicon as viable alternatives to carbon. In fact, silicon is perhaps the most frequent consideration as an alternative to carbon. However, there are problems with that theory as silicon has larger atoms than carbon. In addition, it does not have the ability to form the chemical bonds required for metabolism. Silcone-based chemicals may, however, be more stable in a sulfuric-acid-rich environment such as might be found on some extraterrestrial planets. Another consideration is that silicon dioxide, commonly found in sand, is solid when at a temperature where water remains liquid. So the question becomes if a silicon-based life form also breathes oxygen, like its carbon-based life form, wouldn't it then exhale silicon-dioxide? And if that were true then would not the respiratory system of the silicon-based life form be filled with sand? But what if this silicon-based life form lived where the temperature was hot enough to melt the silicone? Now we come to the exciting possibility that oxygen breathing silicon-based life could exist.

Think of it. There exist planets like Mercury and Venus where such life could exist. And what of our own planet Earth? Do we not have a hot molten center? And what life exists deep inside our volcanoes still active today? Silicon blocks the flow of heat thus it has the potential of protecting a life form built from its particular chemical compound. Consider that diatoms, a major group of algae that are among the most common types of phytoplankton, have a silicate skeletal structure. They use carbon-based proteins for their metabolic needs. So perhaps these are among the life-forms that exist on planets with temperatures and atmospheric pressure different from our Earth. This could substantiate Cairns-Smith's theory that Earth's first living organisms were clay minerals, most likely based on silicon.[14]

Within minerals, clay minerals, the essential constituents of clays, are omnipresent at the earth's surface where microorganisms live, and due to their specific properties they can

[14] http://www.space.com/6456-life-crystal-code.html

interact, positively as a rule, with them. Some clay minerals are being used, either as active principles (gastrointestinal protectors, laxatives, antidiarrhoeaics), or as excipients (inert bases, emulsifiers, lubricants) in certain medicines. Specific properties of clay minerals such as the nanometer size and thin platy or fibrous shapes, the negative electric charge, and high adsorption and absorption capacities justify the therapeutic uses referred to. Also, these and other properties justify the use of clay minerals for improving environmental quality which is necessary for the living quality of man and other organisms as they can act as catalysts for potentially benign chemical processes. "Clay minerals, the most abundant and chemically active components of the surface mineral world, are the key to understanding the links between nature (life), its substrate (essentially silicates), and a mastery of the total ecosystem by man"[15]

The Nitrogen in our DNA
The Calcium in our Teeth
The Iron in our Blood
The Carbon in our Apple Pies
Were Made in the Interiors
of Collapsing Stars
We are Made of Star Stuff

Carl Sagan

The evolution of life from seawater could be supported by the similarity of the compositions of seawater, human body serum, and human red cells, in terms of free cations and anions such as calcium, magnesium, potassium, sodium, chloride, bicarbonate and phosphate (Ulf Lindh, 2005). According to Robert M. Hazen, the investigations carried out on the origin of life lead to the conclusion that minerals must have played key roles in virtually every phase of life emergence, catalyzing the synthesis of key biomolecules, and selecting, protecting and concentrating these molecules. They jump-started metabolism, and may even have acted as first genetic systems.[16]

Consider the fact that silica is in human connective tissue. It is in our bones, our teeth, our glands and organs. It is even in our skin and eyes! Silica is a large part of collagen, the very thing that keeps our skin's elasticity. Silica plays a part in helping calcium keep our bones strong. There are small organisms like plants and sponges which depend on silica for their structural strength. Could silica have played a major role in the building of life on Earth?

Silicon is not only necessary for the formation of the cellular "architecture" that underlies the skeletal system and the connective tissues, it is essential for the functional strength and stability of the tissues as well. A list of all the tissues in the human body that contain high concentrations of silicon gives an idea of how important silicon is to the maintenance of human health, major tissues in the body such as the skeletal framework, the various blood vessels - the aorta in particular, the heart and musculature, the skin and hair, the cartilage and ligaments, as well as soft tissues like the liver, the lungs and the brain.

Silicon is an essential trace mineral needed by our body for stronger bones, physically powerful and supple joints as well as gleaming skin. In the human body, silicon is present in the form of silicic acid or silanate – derived from ether. The presence of silicon in our

[15]] http://us-rem.com/?restore=geological-clay
[16] http://elements.geoscienceworld.org/content/1/3/135.abstract

diet augments the general advantages offered by glucosamine, calcium, in addition to vitamin D.

There are also other essential functions that silicon fulfills in the human body in combination with mucopolysaccharides such as the rapid healing of tissues, chemical regulations involving the transfer of nutrients and water in the membranes of the connective tissues as well as embryonic development and growth.

The term trace mineral is applicable to silicon as it is required in certain processes and structures in the human body. While the mineral silicon probably has a role in the maintenance and formation of major connective tissues, the actual functions of silicon in the body are not yet well understood. Certain areas of the bone that are in the process of mineralization have stores of silicon in large amounts. These indicate that the mineral is vital for the functioning of bones at a normal rate. The actual evidence for this situation in humans is still not available from any research carried out to date. Silicon can be found in many foods, such as honey, whole grain breads, almonds and apples. And silicon is essential for the preservation of cardiovascular health. As far as we know, there are no toxic side effects associated with a high dietary intake of silicon. However, silicosis is a respiratory disease resulting from long term inhalation of large amounts of silicon dust.

According to David Stephenson, a geologist at CalTech, the center of our planet Earth is actually a crystal. He believes that over time this crystal grew until it became an aggregate of many crystals. In the past, the textbook description of the Earth's inner areas such as the mantle and core were shown as fairly simple and similar regions. But with the advances of modern geology, more sophisticated seismographs are being used to gather data about how seismic waves travel through the Earth. The inner core is a solid ball that measures approximately 750 miles in diameter, while its outer core is made up of mostly liquid iron. During the late 1990s geologists noticed that the seismic waves traveling East to West took longer to arrive at their destinations than those traveling on a North to South path. The theory is that perhaps the way the crystals within the Earth's core aligned themselves during formation sets up an alignment known as anisotropy which makes it easier for the seismic waves to travel in that direction.[17] Could further research into this phenomenon lead to discoveries that will permit time travel in the future?

When we consider the origin of life, there are two thoughts to follow. First is water and second is a magnetic field of protection. In the March 5, 2010 issue of Science Daily, a team of researchers stated that Earth's magnetic field was developed by 3.45 billion years ago. During that time period Earth was being blasted with interplanetary debris and the atmosphere was being filled with oxygen. Without the protection of this magnetic field, the solar radiation would cause our water to evaporate, the atmosphere would be destroyed and finally all life would disappear from our surface. Scientists have been working with crystals of quartz from the Greenstone Belt in South Africa to measure the strength of primitive Earth's magnetic field. John Tarduno of the University of Rochester, a co-author of

[17] http://www.astrobio.net/pressrelease/2539/liquid-crystal-life

the study, says that these quartz crystals are the perfect capsule because they are not affected by later events yet have inclusions of iron-containing magnetite in them.

The Rochester research team has found that these magnetite minerals align themselves with the Earth's magnetic field in a cooling magma and then lock into position when the temperature of the quartz goes below 1076°F. What this shows is that the magnetic field would have been much closer to our Earth's atmosphere in its beginnings than it is today. It also suggests that the Earth would lose its lighter elements, such as Hydrogen and Helium, much more rapidly than today. So if you think about it, in today's environment we are losing record volumes of water. Rivers and lakes are drying up, the oceans receding. And yet, the Earth is made up of mostly water. When you consider this, you must realize that the Palaeo-Earth must have been made up almost entirely of water.

The magnetic field is formed by the convection currents of the Earth's mantle. Convection currents are caused by the very hot material at the deepest part of the mantle rising, then cooling, sinking again and then heating, rising and repeating the cycle over and over, pushing the Earth's crust as it moves. Next time you are boiling water, notice how the heat moves the water. This is essentially what happens in the Earth's mantle.

Naturally occurring crystals like Quartz, Sucrose (table sugar), Topaz, Berlinite (a rare phosphate mineral that is structurally identical to Quartz) and Rochelle Salt contain an electric charge known as Piezoelectricity when put under pressure. This was demonstrated by the Curie brothers in 1880 using their knowledge of crystal structure and the piezoelectrical effect studied earlier by Carl Linnaeus, Franz Aepinus, René Just Haüy and Antoine César Becquerel. They found that Quartz and Rochelle Salt showed the most piezoelectricity. Another scientist, Gabriel Lippman, mathematically deduced the reverse effect, which the Curie brothers went on to establish definitive proof of the complete reversibility of electro-elasto-mechanical deformations in piezoelectric crystals. So how does this connect to Earth's real beginning? What is interesting about piezoelectricity is that it is also found in bone and collagen. Research at the University of Pennsylvania in the late 70s and early 80s showed that sustained application of electrical potential could stimulate both resorption and growth (depending on the polarity) of bone in-vivo.[18] Other biological materials with this same ability include tendon, silk, wood, enamel, dentin…and here is the really big one… DNA! There is more, including viral proteins such as those from bacteriophage, which can be used to create a piezoelectric generator strong enough to work a liquid crystal display. [19]

But back to crystals and how they relate to the origins of life on Earth. Crystals by their very nature are periodic, extending regularly in infinite planes, "reproducing" if you will, in three dimensions. Which crystal would be preponderant on primitive earth to possibly do such a job? Silicon dioxide, also known as Silica, is most commonly found in nature as quartz, as well as in various living organisms. Silica is one of the most complex and most abundant families of materials, existing both as several minerals and being produced synthetically. It has been abundant on Earth for billions of years and comes in an amazing variety of allotropes, geometries, and polymorphs. Silicification in and by cells has been common in the biological world for well over a billion years. In the modern world it occurs in bacteria,

[18] http://en.wikipedia.org/wiki/Piezoelectricity

[19] Carlisle, EM (1986). "Silicon as an essential trace element in animal nutrition". Ciba Foundation symposium. Novartis Foundation Symposia 121: 123–39. doi:10.1002/9780470513323.ch8. ISBN 978-0-470-51332-3. PMID 3743227.

Equisetaceae

Dicotyledon

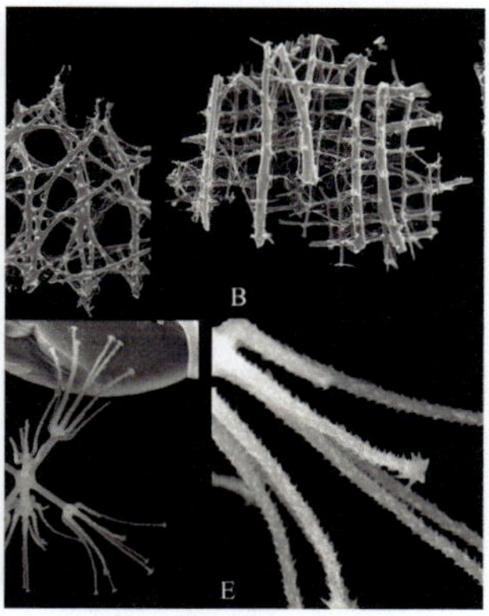

Sponge Spicules

single-celled organisms, plants, and animals (invertebrates and vertebrates). Prominent examples include:

- Tests or frustules of diatoms and Radiolaria.
- Silica phytoliths in the cells of many plants, including Equisetaceae, practically all grasses, and a wide range of dicotyledons.
- The spicules forming the skeletons of many sponges.

As Quartz, it creates more than 10% of the Earth's crust. Thin films of silica grow spontaneously on silicon wafers via thermal oxidation. It is unclear in what ways silica is important in the nutrition of animals. This field of research is challenging because silica is ubiquitous and in most circumstances dissolves in trace quantities only. All the same, it certainly does occur in the living body, leaving us with the problem that it is hard to create proper silica-free controls for purposes of research. This makes it difficult to be sure when the silica present has had operative beneficial effects, and when its presence is coincidental, or even harmful. The current consensus is that it certainly seems important in the growth, strength, and management of many connective tissues. This is true not only for hard connective tissues such as bone and tooth but possibly in the biochemistry of the sub-cellular enzyme-containing structures as well.[20]

Clay minerals are typically formed over long periods of time by the gradual chemical weathering of rocks, usually silicate-bearing, by low concentrations of carbonic acid and other diluted solvents. Clay deposits are typically associated with very low energy depositional environments such as large lakes and marine basins.

Primary clays, also known as kaolins, are located at the site of formation. Secondary clay deposits have been moved by erosion and water from their primary locations.[22]

Belief in the present ongoing spontaneous generation of certain forms of life from non-living matter goes back to Aristotle and ancient Greek philosophy and continued

[20] http://www.astrobio.net/pressrelease/2934/earths-mineral-evolution

Creation Code

Rolando Rivero

$$O = O^{12}\ \Delta^{20}$$

Multiverse
12 + 20 + 1 = 33

Sugar DNA 8+8+8+8=32=1 double helix

Silicon Quartz Crystals 32 + 1 = 33

Bee chromosomes 32 +1=33

to have support in Western scholarship until the 19th century. Classical notions of spontaneous generation, which can be considered under the modern term abiogenesis, held that certain complex, living organisms are generated by decaying organic substances. The first experimental evidence against spontaneous generation came in 1668 when Francesco Redi showed that no maggots appeared in meat when flies were prevented from laying eggs. It was gradually shown that, at least in the case of all the higher and readily visible organisms, the previous sentiment regarding spontaneous generation was false. The alternative seemed to be biogenesis: that every living thing came from a pre-existing living thing (omne vivum ex ovo, Latin for "every living thing from an egg"). By the middle of the 19th century, the theory of biogenesis had accumulated so much evidential support, due to the work of Louis Pasteur and others, that the alternative theory of spontaneous generation had been effectively disproved. Charles Darwin suggested that perhaps the original spark of life could have developed in a warm pond full of ammonia, phosphoric salts, heat, electricity and other elements. He felt that these conditions would have allowed for a protein compound to have been formed that could metamorphose into something more complex.

No new notable research or theory on the subject appeared until 1924, when Alexander Oparin reasoned that atmospheric oxygen prevents the synthesis of certain organic compounds that are necessary building blocks for the evolution of life. In his book *The Origin of Life*, Oparin proposed that the "spontaneous generation of life" that had been attacked by Louis Pasteur did in fact occur once, but was now impossible because the conditions found on the early Earth had changed, and preexisting organisms would immediately consume any spontaneously generated organism. Oparin argued that a "primeval soup" of organic molecules could be created in an oxygenless atmosphere through the action of sunlight. These would combine in ever more complex ways until they formed coacervate droplets. These droplets would "grow" by fusion with other droplets, and "reproduce" through fission into daughter droplets, and so have a primitive metabolism in which those factors which promote "cell integrity" survive, and those that do not become extinct. Many modern theories of the origin of life still take Oparin's ideas as a starting point.

Around the same time, J. B. S. Haldane suggested that the Earth's prebiotic oceans—different from their modern counterparts—would have formed a "hot dilute soup" in which organic compounds could have formed. J.D. Bernal, a pioneer in x-ray crystallography, called this idea biopoiesis or biopoesis, the process of living matter evolving from self-replicating but nonliving molecules, and proposed that biopoiesis passes through a number of intermediate stages.

One of the most important pieces of experimental support for the "soup" theory came in 1952. A graduate student, Stanley Miller, and his professor, Harold Urey, performed an experiment that demonstrated how organic molecules could have spontaneously formed from inorganic precursors, under conditions like those posited by the Oparin-Haldane Hypothesis. The now-famous "Miller–Urey experiment" used a highly reduced mixture of gases—methane, ammonia and hydrogen—to form basic organic monomers, such as amino acids. This provided direct experimental support for the second point of the "soup" theory, and it is around the remaining two points of the theory that much of the debate now centers. In the Miller–Urey experiment, a mixture of water, hydrogen, methane, and ammonia was cycled through an apparatus that delivered electrical sparks to the mixture. After one week, it was found that about 10% to 15% of the carbon in the system was now in the form of a racemic mixture of organic compounds, including amino acids, which are the building blocks of proteins.

The underlying hypothesis held by Oparin, Haldane, Bernal, Miller and Urey was that conditions on the primeval Earth favored chemical reactions that synthesized complex organic compounds from simple precursors. A 2011 reanalysis of the saved vials containing the original extracts that resulted from the Miller and Urey experiments, using current and more advanced analytical equipment and technology, has uncovered more biochemicals than originally discovered in the 1950s. One of the more important findings was 23 amino acids, far more than the five originally found.[21]

Compounds found on meteorites suggest that the chirality of life derives from abiogenic synthesis, since amino acids from meteorites show a left-handed bias, whereas sugars show a predominantly right-handed bias, the same as found in living organisms.[22]

Abiogenesis or biopoiesis is the natural process by which life arose from non-living matter such as simple organic compounds. The earliest life on Earth existed at least 3.5 billion years ago, during the Eoarchean Era when sufficient crust had solidified following the molten Hadean Eon. The earliest specific evidence for life on Earth is biogenic graphite in 3.7 billion-year-old metasedimentary rocks discovered in Western Greenland and microbial mat fossils found in 3.48 billion-year-old sandstone discovered in Western Australia.[23]

The Hadean Earth is thought to have had a secondary atmosphere, formed through degassing of the rocks that accumulated from planetesimal impactors. At first it was thought by scientists like Harold Urey that the earth's atmosphere was made up of hydrides—methane, ammonia and water vapor, and that life began under such reducing conditions, conducive to the formation of organic molecules. However, it is now thought that the early atmosphere, based on today's volcanic evidence, would have contained 60% hydrogen, 20% oxygen (mostly in the form of water vapor), 10% carbon dioxide, 5 to 7% hydrogen sulfide, and smaller amounts of nitrogen, carbon monoxide, free hydrogen, methane and inert gases. As Earth lacked the gravity to hold any molecular hydrogen, this component of the atmosphere was rapidly lost during the Hadean period. Solution of the carbon dioxide in water is thought to have made the seas slightly acidic, with a pH of about 5.5.[24] The atmosphere at the time has been characterized as a "gigantic, productive outdoor chemical

[21] http://en.wikipedia.org/wiki/Abiogenesis

[23] http://en.wikipedia.org/wiki/Clay [24] http://www.astrobio.net/exclusive/2883/diamonds-may-be-lifes-birthstone

l

laboratory."[25] It is similar to the mixture of gases released by volcanoes, which still support some abiotic chemistry today.[26]

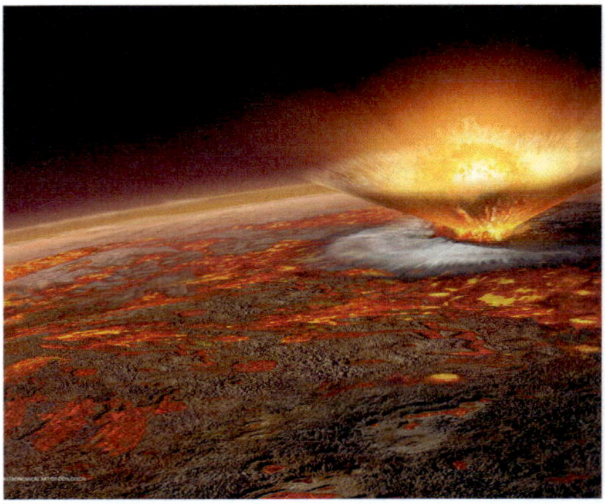

Oceans may have appeared first in the Hadean eon, as soon as two hundred million years (200 Ma) after the Earth was formed, in a hot 100 °C (212 °F) reducing environment, and the pH of about 5.8 rose rapidly towards neutral.[27] This has been supported by the dating of 4.404 Ga-old zircon crystals from metamorphosed quartzite of Mount Narryer in Western Australia, which show that oceans and continental crust existed within 150 Ma of Earth's formation.[28] It has been suggested that between 4.4 and 4.3 Ga, the Earth was a water world, with little if any continental crust, with an extremely turbulent atmosphere and a hydrosphere subject to high UV, from a T Tauri sun and cosmic radiation and continued bolide impact.[29]

By examining the time interval between the devastating environmental events of meteorological bombardment and the constant heating between the Earth's core and the mantle, the time interval when life might first have come into existence can be found for different early environments. A study by Maher and Stevenson shows that if the deep marine hydrothermal setting provides a suitable site for the origin of life, abiogenesis could have happened as early as 4.0 to 4.2 Ga, whereas if it occurred at the surface of the Earth, abiogenesis could only have occurred between 3.7 and 4.0 Ga.[30]

Through the belief that life originated spontaneously on Earth, the Miller–Urey experiment and similar experiments demonstrated that most amino acids, sometimes called "the building blocks of life", can be racemically synthesized under conditions considered similar to those of the early Earth. Lazcano and Miller (1994) suggest that the rapidity of the evolution of life is dictated by the rate of recirculating water through mid-ocean submarine vents. Since complete recirculation takes 10 million years, any organic compounds produced by then would be altered or destroyed by temperatures exceeding 300 °C (572 °F). They estimate that the development of a 100 kilobase genome of a DNA/protein primitive heterotroph into a 7000 gene filamentous cyanobacterium would have required only 7 Ma. Christian de Duve argues that the determination of chemistry means that "life has to emerge quickly...chemical reactions happen quickly or not at all; if any reaction takes a millennium to complete then the chances are all the reagents will simply dissipate or break down in the

[25] http://www.nature.com/news/2004/040220/full/news040216-18.html
[26] http://www.sci-news.com/paleontology/science-fossilized-ecosystems-australia-01537.htm
[27] http://www.astrobio.net/exclusive/2883/diamonds-may-be-lifes-birthstone
[28] http://en.wikipedia.org/wiki/Abiogenesis
[39] http://en.wikipedia.org/wiki/Abiogenesis
[30] http://en.wikipedia.org/wiki/Miller%E2%80%93Urey_experiment

Could cyanobacteria provide the answers to the primordial soup question?

meantime, unless they are replenished by other faster reactions".[31]

There is still no "standard model" of the origin of life. Most currently accepted models draw at least some elements from the framework laid out by Alexander Oparin (in 1924) and John Haldane (in 1925), who postulated the molecular or chemical evolution theory of life.[32] According to them, the first molecules constituting the earliest cells "were synthesized under natural conditions by a slow process of molecular evolution, and these molecules then organized into the first molecular system with properties with biological order."[33] Oparin and Haldane suggested that the atmosphere of the early Earth may have been chemically reducing in nature, composed primarily of methane (CH4), ammonia (NH3), water (H2O), hydrogen sulfide (H2S), carbon dioxide (CO2) or carbon monoxide (CO), and phosphate (PO43-), with molecular oxygen (O2) and ozone (O3) either rare or absent. In such a reducing atmosphere, electrical activity can catalyze the creation of certain basic small molecules (monomers) of life, such as amino acids. This was demonstrated in the Miller–Urey experiment by Stanley L. Miller and Harold C. Urey reported in 1953.

Alexander Graham Cairns-Smith, an organic chemist at the University of Glasgow, sees a significant relationship between the structure of DNA molecules and the structure of certain kinds of mineral crystals. He says that while patterned structures that replicate themselves are common in the inorganic world of crystals, it is a rare quality in the organic world — DNA and RNA are the only organic molecules we know of that strongly exhibit this characteristic.

John Desmond Bernal coined the term biopoiesis in 1949 to refer to the origin of life, and suggested that it occurred in three "stages": 1) the origin of biological monomers; 2) the origin of biological polymers; and 3) the evolution from molecules to cells. In 1949, the Irish scientist J.D. Bernal suggested that clay minerals may have created a meeting place for life's first molecules. Such a scenario could explain how the randomly dispersed molecules of life managed to come together in the diffuse primordial soup.[34]

Cairns-Smith's idea takes Bernal's theory a step further. In his view, clay mineral layers not only attracted certain chemicals from the environment to their surfaces, the mineral layers also acted as the first genetic information carriers, much as the base pairs in DNA do today.

"The objects that I'm particularly interested in are mixed-layered crystals, in which the crystal structure consists of beautifully formed layers packed on top of each other, but with

31] http://en.wikipedia.org/wiki/Abiogenesis
[32] http://en.wikipedia.org/wiki/Formation_and_evolution_of_the_Solar_System
[33] http://en.wikipedia.org/wiki/Abiogenesis
[34] http://www.creationresearch.org/crsq/articles/36/36_4/abiogenesis.html

an arbitrary sequence," says Cairns-Smith. "In that respect, they're like a DNA molecule, which has base pairs, little platelets inside it which are stacked on top of each other. It is the sequence of this stacking which creates the information."

Cairns-Smith doesn't think the clay mineral crystals were "alive" any more than a DNA sample is thought to be alive. Instead, by acting as the first genetic materials for early life, clay mineral crystals created a link between the worlds of inorganic and organic chemistry.

At some point, life launched free of its inorganic genetic origins — the organic substances that evolved from chemical interactions on the mineral layers became stable enough to live apart from their birthplace, and complex enough to replicate themselves into the future.

Some mineral layer combinations probably worked better than others when it came to marshalling the organic molecules that were to eventually become genetic materials. One of his favorite contenders for life's early mineral template is authigenic chlorite, which can create complicated shapes that resemble brussels sprouts. Such chlorite crystals growing inside sandstone often coat the sand grains and do not block the flow of solutions within the rock — a potentially important quality for the very first evolving systems. However, rather than one particular mineral layer sequence leading to life, Cairns-Smith thinks many different mixed layer structures might have contributed to life's evolution.

Cairns-Smith first developed his theory in the sixties and promoted it through the publication of several books and articles. Many scientists found the idea intriguing; however, others discounted it because life today shows no evidence of possessing inorganic origins. The majority of scientists who work on the problem of life's origin instead focus on how simple sugars, amino acids, and other organic chemicals came together to form nucleic acids and proteins.

Cairns-Smith believes such a chain of events was improbable on the early Earth — the nucleic acid and protein system of life is too complex to have sprung outright from simple ingredients. Even the RNA world hypothesis, which envisions RNA playing dual roles that today are carried out by DNA and proteins, is a relatively advanced and sophisticated process.

"A simpler kind of evolution came first, and then what are now the molecules of life came to be produced in a consistent way," says Cairns-Smith. "Of course there was no foresight here, but as soon as an evolutionary process was underway, the world would have changed and nature would have had a new set of toys to play with."[35]

If crystals provide the order that life needs to arise, that has implications for the possibility for life beyond Earth. Mineral crystals should be common on rocky worlds, even if those planets have chemical environments quite different from our own. If Cairns-Smith's theory

[35] http://www.newworldencyclopedia.org/entry/Origin_of_life

is correct, then the spark of life may be shimmering on crystal surfaces throughout the universe.[34]

While features of self-organization and self-replication are often considered the hallmark of living systems, there are many instances of abiotic molecules exhibiting such characteristics under proper conditions. For example, Martin and Russell show that physical compartmentation by cell membranes from the environment and self-organization of self-contained redox reactions are the most conserved attributes of living things, and they argue therefore that inorganic matter with such attributes would be life's most likely last common ancestor. Later, Palasek showed that self-assembly of RNA molecules can occur spontaneously due to physical factors in hydrothermal vents.[36]

The Nov. 23, 2007 issue of Science Magazine featured an article written by Professor Noel Clark, Michi Nakata and Christopher Jones from CU-Boulder, Giuliano Zanchetta and Tommaso Bellini of the University of Milan, Brandon Chapman and Ronald Pindak of Brookhaven National Laboratory and Julie Cross of Argonne National Laboratory, in which they elaborated on findings that some unexpected forms of liquid crystals of ultrashort DNA molecules which, when immersed in water, provided a new scenario for a key step in the emergence of life on Earth.

CU-Boulder physics Professor Noel Clark said the team found that surprisingly short segments of DNA, life's molecular carrier of genetic information, could assemble into several distinct liquid crystal phases that "self-orient" parallel to one another and stack into columns when placed in a water solution. Life is widely believed to have emerged as segments of DNA or RNA-like molecules in a prebiotic "soup" solution of ancient organic molecules.

Since the formation of molecular chains as uniform as DNA by random chemistry is essentially impossible, Clark said, scientists have been seeking effective ways for simple molecules to spontaneously self-select, "chain-up" and self-replicate. The new study shows that in a mixture of tiny fragments of DNA, those molecules capable of forming liquid crystals selectively condense into droplets in which conditions are favorable for them to be chemically linked into longer molecules with enhanced liquid crystal-forming tendencies, he said.

"We found that even tiny fragments of double helix DNA can spontaneously self-assemble into columns that contain many molecules," Clark said. "Our vision is that from the collection of ancient molecules, short RNA pieces or some structurally related precursor emerged as the molecular fragments most capable of condensing into liquid crystal droplets, selectively developing into long molecules."

Liquid crystals - organic materials related to soap that exhibit both solid and liquid properties - are commonly used for information displays in computers, flat-panel televisions, cell phones, calculators and watches. Most liquid crystal phase molecules are rod-shaped

[36] http://www.space.com/6456-life-crystal-code.html

and have the ability to spontaneously form large domains of a common orientation, which makes them particularly sensitive to stimuli like changes in temperature or applied voltage.

RNA and DNA are chain-like polymers with side groups known as nucleotides, or bases, that selectively adhere only to specific bases on a second chain. Matching, or complementary base sequences enable the chains to pair up and form the widely recognized double helix structure. Genetic information is encoded in sequences of thousands to millions of bases along the chains, which can be microns to millimeters in length.

Such DNA polynucleotides had previously been shown to organize into liquid crystal phases in which the chains spontaneously oriented parallel to each other, he said. Researchers understand the liquid crystal organization to be a result of DNA's elongated molecular shape, making parallel alignment easier, much like spaghetti thrown in a box and shaken would be prone to line up in parallel, Clark said.

The CU-Boulder and University of Milan team began a series of experiments to see how short the DNA segments could be and still show liquid crystal ordering, said Clark. The team found that even a DNA segment as short as six bases, when paired with a complementary segment that together measured just two nanometers long and two nanometers in diameter, could still assemble itself into the liquid crystal phases, in spite of having almost no elongation in shape.

Structural analysis of the liquid crystal phases showed that they appeared because such short DNA duplex pairs were able to stick together "end-to-end," forming rod-shaped aggregates that could then behave like much longer segments of DNA. The sticking was a result of small, oily patches found on the ends of the short DNA segments that help them adhere to each other in a reversible way – much like magnetic buttons – as they expelled water in between them, Clark said.

All life on Earth has a similar DNA-based chemistry. Understanding how this chemistry works and how it originated in living systems is important in determining whether or not life could develop on other planets.

A key characterization technique employed was X-ray microbeam diffraction combined with in-situ optical microscopy, carried out with researchers from Argonne and Brookhaven National Laboratories. The team used a machine called the Argonne Advanced Photon Source synchrotron that enabled probing of the "nano DNA" molecular organization in single liquid crystal orientation domains only a few microns in size. The experiments provided direct evidence for the columnar stacking of the nano DNA pieces in a fluid liquid crystal phase.

"The key observation with respect to early life is that this aggregation of nano DNA strands is possible only if they form duplexes," Clark said. "In a sample of chains in which the bases don't match and the chains can't form helical duplexes, we did not observe liquid crystal ordering."

Subsequent tests by the team involved mixed solutions of complementary and non-complementary DNA segments, said Clark. The results indicated that essentially all of the complementary DNA bits condensed out in the form of liquid crystal droplets, physically separating them from the non-complementary DNA segments.

"We found this to be a remarkable result," Clark said. "It means that small molecules with the ability to pair up the right way can seek each other out and collect together into drops that are internally self-organized to facilitate the growth of larger pairable molecules.[37]

Research shows that minerals on Earth have also evolved alongside living organisms – and the diversity of minerals on our planet can be linked to the processes of life.

Evolution isn't just for living organisms. Scientists at the Carnegie Institution have found that the mineral kingdom co-evolved with life, and that up to two thirds of the more than 4,000 known types of minerals on Earth can be directly or indirectly linked to biological activity. The finding, published in the 2008 American Mineralogist, could aid scientists in the search for life on other planets.[38]

Robert Hazen and Dominic Papineau of the Carnegie Institution's Geophysical Laboratory, with six colleagues, reviewed the physical, chemical, and biological processes that gradually transformed about a dozen different primordial minerals in ancient interstellar dust grains to the thousands of mineral species on the present-day Earth. (Unlike biological species, each mineral species is defined by its characteristic chemical makeup and crystal structure.)

"It's a different way of looking at minerals from more traditional approaches," says Hazen." Mineral evolution is obviously different from Darwinian evolution - minerals don't mutate, reproduce or compete like living organisms. But we found both the variety and relative abundances of minerals have changed dramatically over more than 4.5 billion years of Earth's history."

All the chemical elements were present from the start in the solar system's primordial dust, but they formed comparatively few minerals. Only after large bodies such as the sun and planets congealed did there exist the extremes of temperature and pressure required to forge a large diversity of mineral species. Many elements were also too dispersed in the original dust clouds to be able to solidify into mineral crystals.

The rod-shaped D. audaxviator was recovered from thousands of liters of water collected deep in the Mponeng Mine in South Africa. Microbes can survive by interacting with minerals deep below the surface of Earth.

[37] http://www.ehow.com/facts_7496890_information-quartz-crystals.html
[38] http://en.wikipedia.org/wiki/Mineral

As the solar system took shape through "gravitational clumping" of small, undifferentiated bodies - fragments of which are found today in the form of meteorites - about 60 different minerals made their appearance. Larger, planet-sized bodies, especially those with volcanic activity and bearing significant amounts of water, could have given rise to several hundred new mineral species. Mars and Venus, which Hazen and coworkers estimate to have at least 500 different mineral species in their surface rocks, appear to have reached this stage in their mineral evolution.[39]

However, only on Earth – at least in our solar system – did mineral evolution progress to the next stages. A key factor was the churning of the planet's interior by plate tectonics, the process that drives the slow shifting continents and ocean basins over geological time. Unique to Earth, plate tectonics created new kinds of physical and chemical environments where minerals could form, and thereby boosted mineral diversity to more than a thousand types.

What ultimately had the biggest impact on mineral evolution, however, was the origin of life, approximately 4 billion years ago. "Of the approximately 4,300 known mineral species on Earth, perhaps two-thirds of them are biologically mediated," says Hazen. "This is principally a consequence of our oxygen-rich atmosphere, which is a product of photosynthesis by microscopic algae." Many important minerals are oxidized weathering products, including ores of iron, copper and many other metals [40].

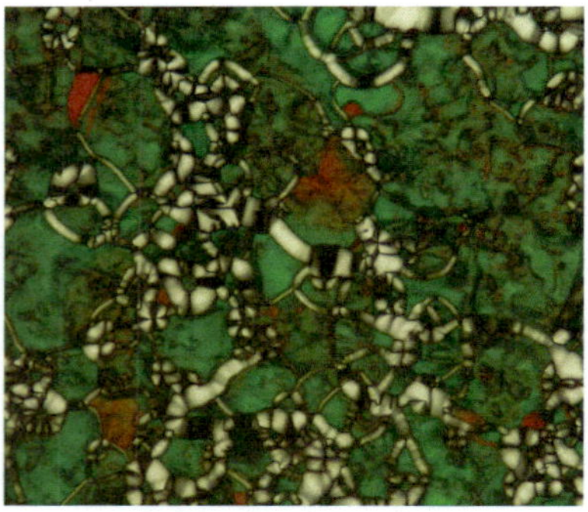

Credit: Michi Nakata
University of Colorado at Boulder

Cyanobacteria is a type of bacteria that get their energy through photosynthesis and are thought to have been the first microbes to produce oxygen by photosynthesis. By producing oxygen as a gas as a by-product of photosynthesis, cyanobacteria are thought to have converted the early reducing atmosphere into an oxidizing one, thus dramatically changing the composition of life forms on Earth by stimulating biodiversity and causing the near-extinction of oxygen-intolerant organisms. According to endosymbiotic theory, the chloroplasts found in plants and eukaryotic algae may have evolved from cyanobacterial ancestors via endosymbiosis. Cyanobacteria are a photosynthetic nitrogen fixing group that survive in a wide variety of habitat, soil and water. People living near fresh water or other marine environments have most likely seen the highly visible blue-green algae bloom at some point. It is a common occurrence in the tributaries of the St. Johns river, here in Jacksonville, Florida.

Microorganisms and plants also accelerated the production of diverse clay minerals. In the oceans, the evolution of organisms with shells and mineralized skeletons generated thick, layered deposits of minerals such as calcite, which would be rare on a lifeless planet.

[39] http://www.sciencedaily.com/releases/2008/11/081113181035.htm
[40] http://carnegiescience.edu/news/mineral_kingdom_has_co_evolved_life

"For at least 2.5 billion years, and possibly since the emergence of life, Earth's mineralogy has evolved in parallel with biology," says Hazen. "One implication of this finding is that remote observations of the mineralogy of other moons and planets may provide crucial evidence for biological influences beyond Earth."

Stanford University geologist Gary Ernst called the study "breathtaking," saying that "the unique perspective presented in this paper may revolutionize the way Earth scientists regard minerals."[41]

One of the hurdles in origin of life theories is that the pieces that make up complex biomolecules do not readily come together by themselves. A group of scientists proposes that diamonds provided a kind of "work bench" for biomolecule manufacturing on early Earth.

WHAT IS MOST IMPORTANT? Us or the habitat? Who put the first seed? Where did the conifers of early earth come from? Not long after its formation, our planet was – according to astrobiologists – awash in a primordial soup that contained the rudimentary ingredients of life. The fly in the "soup" theory, however, is that the small molecular bits likely needed outside help in order to latch together into the long, complex biomolecules that living organisms use.

Some scientists have suggested that the surfaces of minerals on the early Earth provided an organizing platform upon which the building blocks of life could assemble. Recent studies of diamond suggest that its surface would be especially good for this.

"Diamond is totally non-toxic, an excellent biomaterial and certainly the only naturally existing material that is completely biocompatible on all levels – presumably the best of all possible platforms for the formation of life," said Andrei Sommer from the University of Ulm in Germany. [42]

Sommer and his colleagues discovered that a certain type of diamond, called hydrogenated diamond, imposes a rigid order on molecules near its surface. They suggest, in a 2008 issue of the journal Crystal Growth and Design, that this diamond-mediated order helped fit together the pieces that led to the emergence of life.

Using a humidity chamber, researchers were able to detect the formation of crystal water layers on a hydrogenated diamond through a measured decrease in electrical conductivity in response to an increase in humidity.

Hydrogenated diamonds are just diamonds with an outer coating of hydrogen atoms, but they are not something you'll find in your local jewelry shop. In fact, the only hydrogenated diamonds currently known are all made in the lab.

"In nature, diamond hydrogenation is likely to occur in or in the vicinity of volcanoes known to emit a variety of hot gases including hydrogen," Sommer said. The early Earth had

[41] https://www.gl.ciw.edu/news/mineralogy/mineral_kingdom_has_coevolved_with_life

[42] Cryst. Growth Des., 2008, 8 (8), pp 2628–2629. DOI: 10.1021/cg8005037. Publication Date (Web): July 10, 2008. Copyright © 2008 American Chemical Society

so much volcanic activity that he thinks it is highly probable that hydrogenated diamonds existed back then.

Sommer and his collaborators previously showed that hydrogenated diamond is very hydrophobic, or "water fearing" -- meaning it pushes water away. When hydrogenated diamond is wet, the water molecules line up on the surface as if they were frozen into a crystal layer (an analogy might be static electricity making all the hairs on your head point out).

Surprisingly, these crystal water layers do not disappear when the hydrogenated diamond is fully immersed in water. Because this is the only natural material known to exhibit this behavior, Sommer's team proposes that small organic molecules in the primordial soup landed on hydrogenated diamond and were helped by its robust crystal water layers into linking together to form proteins and DNA.[43]

Support for this idea comes from a recent study that found that certain nucleobases (the building blocks of DNA and RNA) form an organized pattern on the surface of graphite, which is chemically similar to diamond.

Hydrogenated diamond should be a better organizing platform than graphite, Sommer said. This is because the crystal layers that form on it are not static; they change with temperature and light intensity. The resulting fluctuations could have helped drive the development of novel molecules in the primordial soup.

It is believed that the early Earth experienced a great deal more volcanic activity than it does today. If so, then diamond hydrogenation, which occurs in or in the vicinity of volcanoes, may have been quite common.

"So far, crystal water layers have only been described on hydrogenated diamond," said Horst-Dieter Foersterling of Philipps University of Marburg, who was not involved with this work. "This is a new field of research. That this system can be helpful for the formation of biomolecules is a plausible hypothesis."

It remains uncertain whether there was any hydrogenated diamond on Earth billions of years ago, but even a little bit might be enough.

It is important to note that silicon sits just below carbon on the periodic table. Due to its larger atomic radius, silicon does not form double bonds with the oxygen atoms in SiO2. This means that each silicon atom in a quartz crystal is at the center of a tetrahedron of oxygen atoms, and that each oxygen atom shares two silicon atoms with its neighbors. The structure of quartz crystals is complex. The SiO4 units are linked into helical chains that twist around one another much like our own DNA. However, when heated above 1500° C, the quartz structure will change. Each silicon atom will be at the center of a tetrahedron of silicon atoms but with an oxygen atom in between each pair of silicons.

[43] http://www.colorado.edu/news/releases/2007/11/22/tiny-dna-molecules-show-liquid-crystal-phases-pointing-new-scenario-first com/releases/2008/11/081113181035.htm

Silicon Quartz Crystal Creation Code

Quartz belongs to the trigonal crystal system. The ideal crystal shape is a six-sided prism terminating with six-sided pyramids at each end. In nature, quartz crystals are often twinned, distorted, or so intertwined with adjacent crystals of quartz or other minerals as to only show part of this shape or to lack obvious crystal faces altogether and appear massive. Well-formed crystals typically form in a "bed" that has unconstrained growth into a void; usually the crystals are attached at the other end to a matrix and one termination pyramid is present. A Quartz geode is such a situation; the void is approximately spherical in shape and lined with a bed of crystals pointing inward.

Quartz is a mineral and is different than a rock due to its ordered atomic structure and the fact that it has a specific chemical compound usually abiogenic, unlike rocks which can be an aggregate of minerals or non-minerals and do not have a specific chemical composition. There are nearly 5,000 known mineral species. Of those, silicate minerals make up over 90% of the Earth's crust! Silicon and oxygen make up 75% of the Earth's crust which means that the Earth consists mainly of silicate minerals. Eight elements account for most of the key components of minerals, due to their abundance in the Earth's crust. The eight elements in order of decreasing abundance are oxygen, silicon, aluminum, iron, magnesium, calcium, sodium and potassium. By weight, 46% of the Earth's crust is oxygen while 28% of its weight is made up of silicon, making these two elements the most important of the eight. Other important mineral groups include the native elements: sulfides, oxides, halides, carbonates, sulfates and phosphates.

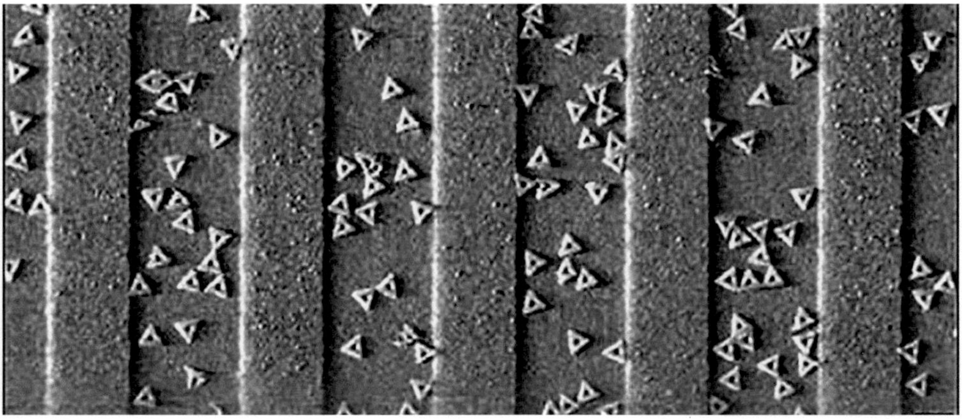

Recent advances in high-resolution genetic and x-ray absorption spectroscopy have shown that microorganisms exist on every rock, soil and particle surface spanning the globe reaching depths at 1600 meters below the ocean floor and 70 kilometers into the stratosphere, possibly entering the mesosphere. Bacteria have contributed to the formation of minerals for billions of years and biologists and geologists are beginning to appreciate the magnitude of mineral geoengineering these microorganisms are capable of. These creatures can precipitate metals from solution contributing to the formation of ore deposits in addition to their ability to catalyze mineral dissolution, to respire, precipitate, and form minerals. In 2009, researchers reported in the scientific journal *Nature Nanotechnology* that they were able to get engineered "DNA origami" to self-organize on silicon creating triangular shapes.[44]

Some say 32 points and 1=33. Some say 32 and 1 are the same. Crystals are restricted to 32 point groups which differ by their symmetry. Due to this 32 point restriction, minerals with different chemistry can have identical crystal structure. These point groups are further grouped into more broad categories. The six crystal families are Isometric, Tetragonal, Orthorhombic, Hexagonal, Monoclinic and Triclinic. Quartz belongs to the Hexagonal family as does calcite and tourmaline.

Creation Code

Rolando Rivero

$$O = O^{12}\ \Delta^{20}$$

Multiverse
12 + 20 + 1 = 33

Sugar DNA 8+8+8+8=32=1 double helix

Silicon Quartz Crystals 32 + 1 = 33

Bee chromosomes 32 +1=33

[44] http://www.caltech.edu/content/caltech-and-ibm-scientists-use-self-assembled-dna-scaffolding-build-tiny-circuit-boards

Bee Chromosomes Creation Code

In 2006, researchers at Oregon State University discovered a bee species that was 100 million years old embedded in an amber specimen from northern Myanmar.[45] This is 35 to 45 million years older than the other known bee fossils. Pollen grains are visible on its back legs. The Melittosphex burmensis, as it is named, has characteristics of both wasps and bees, leading scientists to believe that it may have been alive at a time when the wasp and bee were separating onto their own evolutionary paths. What is interesting is that this may provide a clue as to how the rapid expansion and diversity of flowering plant life came about during the Cretaceous Period. Before this time, the Earth's plant life was mainly covered by conifer trees which depend on wind for reproduction. With the evolution of bees, angiosperm type plants (those requiring means other than wind to spread seeds) began to give rise. Bees are found on every continent in the world except Antarctica.[46] And they are found in every habitat on Earth that contains insect-pollinated flowering plants. Flowering plants are important to the evolution of life because they can grow quickly and spread into new areas more rapidly. And they can also develop into more diverse types of plants.

"[Hornets and wasps] are devoid of the extraordinary features which characterize bees; this we should expect, for they have nothing-divine about them as the bees have."

- Aristotle

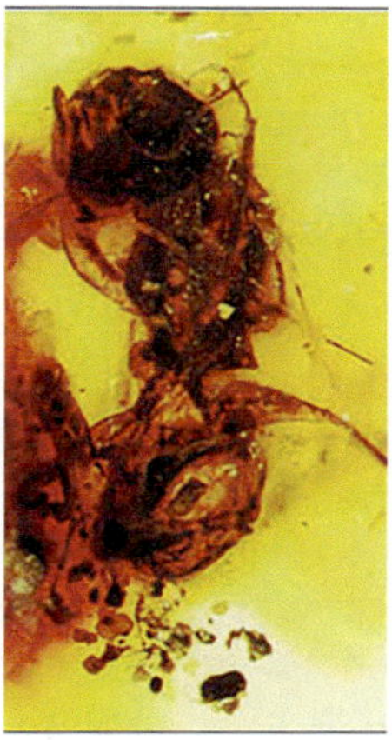

Bees play an important role in pollinating flowering plants, and are the major type of pollinator in ecosystems that contain flowering plants. Flowering plants account for nearly all the food plants on Earth. That means that bees are critical to the very existence of life because it is their act of pollination that supplies food for humans and other animal life. Research by CSIRO entomology scientists has proven that one of every three mouthfuls of food comes from insect pollinated crops.[47] Beekeepers in many countries now contract their bees out for the purposes of pollination of crops. Monoculture and the massive decline of many bee species (both wild and domesticated) has caused many honey bee keepers to become migratory so that bees can be concentrated in seasonally varying high-demand areas of pollination. Insecticides used on blooming plants kill many bees, both by direct poisoning and by contamination of their food supply. A honey bee queen may lay 2000

[45] Deborah R. Smith, Lynn Villafuerte, Gard Otisc & Michael R. Palmer (2000). "Biogeography of Apis cerana F. and A. nigrocincta Smith: insights from mtDNA studies" (PDF). Apidologie 31 (2): 265–279. doi:10.1051/apido:2000121.

[46] Michael S. Engel, I. A. Hinojosa-Diaz & A. P. Rasnitsyn (2009). "A honey bee from the Miocene of Nevada and the biogeography of Apis (Hymenoptera: Apidae: Apini)". Proceedings of the California Academy of Sciences 60 (3): 23–38.

[47] www.terradaily.com/reports/research_discovers_oldest_bee_evolutionary- link Elekonich, M. 2008. Biomedical Research with Honey Bees. 129 – 138. In P. M. Conn (ed.), Sourcebook of Models for Biomedical Research, Humana Press, New Jersey. http://www.sbcs.qmul.ac.uk/staff/larschittka.html

eggs per day during spring buildup, but she also must lay 1000 to 1500 eggs per day during the foraging season, mostly to replace daily casualties, most of which are workers dying of old age. Among solitary and primitively social bees, however, lifetime reproduction is among the lowest of all insects, as it is common for females of such species to produce fewer than 25 offspring. [48]

Further, they have suggested that the honey bee is the first insect to have a system much like that of humans. The Female workers and Queen bees have 32 chromosomes, 16 are contributed by the queen's egg and 16 by the drone's sperm.[49] Drones have only one set of chromosomes. It takes 32 chromosomes to create 1 new Queen bee. Telomeres are the protectors of our chromosomes. The enzyme, telomerase is responsible for replenishing the telomeres in the early development of humans. When the telomerase activity ceases during the aging process, it shortens the telomere. In cancer cells, these telomerase no longer shut down causing unlimited cell division, therefore growing and spreading the cancer. Telomerase activity is also imperative for cloned animals' very survival.

Creation Code

Rolando Rivero

$$O = O^{12}\ \Delta^{20}$$

Multiverse
12 + 20 + 1 = 33

Sugar DNA 8+8+8+8=32=1 double helix

Silicon Quartz Crystals 32 + 1 = 33

Bee chromosomes 32 +1=33

Most researchers believe the honey bee originated in Africa. Bee mythology can be found in ancient cultures where honey was hunted and bees were considered a gift from the gods. The San Bushmen of Kalahari, an ancient hunter-gatherer society from the African Kalahari Desert, tell this creation story:

Bee was a kindly creature. One night, Mantis needed to cross a great, flooded river to reach his family, and asked Bee for help. Bee offered to carry Mantis on her back. She flew over the raging waters but was beaten down by a strong wind. Bee was dangerously close to the waves when she saw a magnificent flower floating on the water. Bee dropped Mantis onto the flower, fell beside her, and died of exhaustion. When the sun rose, curled up on the flower was the first human being.[50]

Numerous Greek myths include bees. They were associated with Demeter, the goddess of farming and the earth, as well as with Apollo, Zeus and Orpheus. Honey was considered the nectar of the gods and symbolized knowledge and wisdom. In the Hindu religion, bees symbolize the spirit of man becoming full of knowledge and is also associated with the bliss of nirvana. The gods Vishnu, Indra and Krishna have been shown as bees sitting on lotus flowers. Ancient Egyptians associated the bee with royalty and power, believing that the bees originated from the tears of their god Ra. And they buried honey with the dead. Both the Atharva Veda and the ancient Greeks associated lips anointed with honey with the gift of eloquence and even of prescience. The priestess at Delphi was the "Delphic Bee". The

[48] http://www.csiro.au/Portals/Media/2006/HoneyBeeGenome.aspx
[49] http://www.glenn-apiaries.com/principles.html
[50] http://www.raymondhuber.co.nz/honey-bees-ebook-life-cycle-honey-stings-crisis-myths/bee-book-download/bee-mythology/

Quran has a chapter titled "The Bee". A community of honey bees has often been employed throughout history by political theorists as a model of human society: this concept appears in the writings of Aristotle and Plato; in Virgil and Seneca; in Erasmus and Shakespeare; in Marx and Tolstoy.

Honey bees, signifying immortality and resurrection, were royal emblems of the Merovingians, revived by Napoleon. And is the heraldic emblem of the Barberini. Honey bees appear to have their center of origin in South and Southeast Asia (including the Philippines), as all but one (i.e. Apis mellifera), of the extant species are native to that region. Notably, living representatives of the earliest lineages to diverge (Apis florea and Apis andreniformis) have their center of origin there.[51]

The first Apis bees appear in the fossil record at the Eocene–Oligocene boundary (23–56 Mya), in European deposits. The origin of these prehistoric honey bees does not necessarily indicate Europe as the place of origin of the genus, only that it occurred there then. A few fossil deposits are known from South Asia, the suspected region of honey bee origin, and fewer still have been thoroughly studied.

No Apis species existed in the New World during human times before the introduction of A. mellifera by Europeans. Only one fossil species is documented from the New World, Apis nearctica, known from a single 14-million-year old specimen from Nevada.[52]

The close relatives of modern honey bees—e.g. bumblebees and stingless bees—are also social to some degree, and social behavior seems a plesiomorphic trait that predates the origin of the genus. Among the extant members of Apis, the more basal species make single, exposed combs, while the more recently evolved species nest in cavities and have multiple combs, which has greatly facilitated their domestication.

Most species have historically been cultured or at least exploited for honey and beeswax by humans indigenous to their native ranges. Only two of these species have been truly domesticated, at least one (A. mellifera) since the time of the building of the Egyptian pyramids, and only that species has been moved extensively beyond its native range.

In his book, "The Life of the Bee", Maurice Maeterlinck popularized the belief that the tiny bee was a vital ecological mainstay, suggesting that we owe our civilization to the bee. And that without it we would see a collapse of civilization. Hundreds of thousands of plants are dependent on the activities of bees. A study by the New York State Department of Agriculture in 1907 indicated that many crops are dependent on bees for proper pollination.

Researchers led by Andrew Feinberg of Johns Hopkins University in Baltimore, Maryland, and Gro Amdam of Arizona State University in Tempe have discovered that even though the genes of honey bees in a hive may have nearly identical genes, they can change the DNA that determines what job function they perform within the hive social structure. What is

[51] http://www.westcoastbeekeepers.com/aboutbees.html
[52] http://cdn.worldheritage.org/articles/Honeybee

even more interesting is that their research also indicates that the bee is able to revert to its original job function. Further research may provide answers to what actually makes this epigenic change occurs.

Could bees possess unusual intelligence? Many processes such as memory retention, visual awareness and spatial reasoning, thought to be exclusively human characteristics, have been seen in bee research. Honey bees display perceptual, cognitive and neurological

capabilities not often observed in other insects. And they have an extreme commitment to the preservation of their complex yet cooperative communities. They must figure out how to survive in all kinds of conditions, relying on individual bees to each do their share to find food, bring it back to the hive and to share its location with others in the hive.

Bees are one of the few species who act individually all for the benefit of the whole group. Very recently there have been discoveries of individual bees' ability to communicate. When a female bee scout has discovered a new location for the hive, she will do a special dance. The dance even includes a special code telling her hive mates the specific location! Experiments have been performed in which a bowl of sugar water was placed in a different location for several days. Each day the bees were able to find the bowl. After several days, the bees were actually waiting at the next location! They had been able to calculate where the next spot would be! "How is that even possible?", you might ask. Apparently bees can solve math problems that humans rely on computers to do for them. They can determine the shortest and most efficient way to hundreds of flowers every day. Not only can they figure this out, they are also able to remember the details of their adventures for trips as long as six miles over days. Further research has shown that individual bees have the ability to understand abstract concepts related to finding food. Could bees have individual intelligence? Humans live in large societies with a collective culture and still maintain their individual intelligence. Could this be true of bees as well? After all, it has been established the genome of the honey bee is very similar to that of humans. (Eleknoich, 2008) Perhaps that study of bees can lead to insight into our own functioning.

The human brain, weighing between 1.25kg and 1.45kg contains an estimated 85 billion nerve cells while the honey bee contains a million nerve cells in its tiny 1 milligram of a brain. Bigger brains might mean bigger hard drives; not necessarily better processors according to Lars Chittka, Professor of Sensory and Behavioral Ecology at Queen Mary's Research Center for Psychology, whose research has established links between sensory physiology and learning psychology on the one hand, and evolutionary ecology on the other. [53]

One of the most fantastic qualities of the bee is its ability to make perfect hexagonal comb cells. It is totally amazing and curious that individual bees each start at their own point within the hive, yet they build the combs' cells exactly the same size! What's more curious is that the junction of the individual cells is invisible and that there is never an error in the angles of their hexagons. Of further interest is the fact that the combs are filled in a specific order. Bees place the eggs in the bottom of the cell. Next is a layer of pollen, then the top half of the comb is filled with honey. There is also always a store of honey around the perimeter of the hive with several rows of pollen filled cells insulating the larval chambers. This protective order not only prevents the three contents from getting mixed up in the darkness of the hive, it also permits humans to remove honey without destroying the new bee larvae. This is a conscious behavior. The queen never mistakenly lays her eggs in empty cells intended for honey or pollen, but always lays them in the correct place. Is this is a discriminating ability given to her by God? How does she know exactly where to place her eggs?

The construction of the comb is dependent on the existence of wax. The fact that this substance is produced by bees is in itself a proof of creation. Many evolutionists suggest that bees did not have this ability at the time they first came into existence, and that all their attributes and behaviors came into being over time, as the result of a series of coincidences. So how did bees discover the ingredients of beeswax—which consists of a substance completely unknown to them? How is it that every single bee has been able to employ the same formula for millions of years? How did bees manage to form the glands and organs necessary to produce such a perfect material as beeswax within their own bodies? And how do the bees know their place within the hive social structure? They must have a conscious intelligence of some sort to be able to know these things.

Neither the bee nor any other living thing on Earth can add new organs to its body at will, nor make them produce entirely new secretions. The physical structures and miraculous abilities in bees clearly prove that they were brought into existence by a Creator. Like all other living things on Earth, bees were created by God, Who manifests peerless examples of His intellect in bees, in order that humans should think about and learn from them. It is extremely difficult for a human to draw regular geometrical shapes without such implements as a ruler and set square. It is nearly impossible to get the 120-degree internal

[53] http://www.sciencedaily.com/releases/2009/11/091117124009.htm

angles of a hexagon right, as bees manage to do—in near-total darkness. How is it that bees have been able to construct combs of this same perfection for millions of years? Bees have no ability to calculate angles; neither have they any knowledge of geometrical shapes. It is God, Who created the entire universe, Who inspires in bees the knowledge and ability with which they can build their combs.

Most of the tasks performed by bees are quite astonishing to us. During a bee's short life span of only a few weeks, he performs all the jobs in the hive in a specific order, from caring for the young to construction, from food gathering to honey production, succeeding at every step.

The bee's nervous system, which enables all these jobs, consists of around 7,000 neurons. A human being has some 2 million times that number. However, as we have seen in detail, bees are able to carry out all these tasks to perfection:

-They perform a series of complicated tasks in the hive, such as feeding the larvae, cleaning, air conditioning, maintenance and repair.

-They are able to distinguish between friendly and hostile bees.

-They are able to give directions according to the angle of the Sun.

-They are able to perceive ultraviolet rays.

-They are able to calculate the weight of the pollen they are carrying.

-They are able to carry out in-flight corrections to their course by looking at the brightness of the sky and landmarks and perceiving odors along their route.

-They are able to calculate the distance they have covered in flight.

-They can measure the frequency of the movements in the dance performed in the hive and thus calculate the distance of the food source.

-Even though the dance takes place in a vertical plane, they are able to accurately calculate the angle between the Sun and the food source.

-They are able to build exceptionally regular hexagonal comb cells.

-Bees will not only avoid toxic plants (including those sprayed with harmful pesticides), but they also seek those plants that contain the highest nutritional values. Bee pollen is one of the richest and purest natural foods ever discovered, and the incredible nutritional and medicinal value of pollen has been known for centuries.

Pollen grains contain the male germ cells (elements) that are produced by all plants, flowers or blossoms. This is essential in order to ensure that plant life throughout the world continues by a process involving fertilization and plant embryo formation.

One teaspoonful of pollen contains approximately 1,200 pellets or 2.5 billion grains, each of which has the capacity to supply those factors that are necessary in order to fertilize and reproduce the particular species that it represents (such as a fruit, grain or tree). Pollen is composed of myriads of micro spores that are produced in the anthers of flowers and in the cones of conifers.

Many plants are pollinated by wind, rain or water-currents, while colorfully attractive or scented flowers containing nectar are largely pollinated by insects (including flies, bees, wasps, butterflies, beetles and moths).

Pollen gathered by bees is superior to that obtained directly from flowering plants. The bees are extremely discriminate about selecting the best pollen from the millions of grains that are present. Of these, only two types are found, namely, anemophile pollen grains (which are not collected by bees, and produce allergic reactions) and entomophile pollen grains (which are collected by bees, and possess greater nutrient content). In actuality, entomophile pollen grains have been employed in the successful treatment of airborne pollen allergies. It is apparent that the bees only select those grains of pollen that are rich iin all the nutrients, especially nitrogenous materials. The bees mix the pollen grains with a sticky substance that is secreted from their stomachs, which allows the pollen to adhere to their rear legs in "pollen baskets" in order to safely transport it to their hives.

Creation Code

Rolando Rivero

$$O = O^{12}\,\Delta^{20}$$

Multiverse
12 + 20 + 1 = 33

Sugar DNA 8+8+8+8=32=1 double helix

Silicon Quartz Crystals 32 + 1 = 33

Bee chromosomes 32 +1=33

Many other flowers are also pollinated by certain birds, such as sun birds, honey creepers, lorikeets and hummingbirds. Marsupials (such as honey "mice" and bats) will also pollinate certain flowering plants, and even snails have been observed transporting pollen.

Pollens are usually designated by their flower origin in order to establish certain preferences that are dependable. The color and shape usually indicates the species of plant from which it was obtained, as well as the specific geographical region. Although the color of pollen is normally unimportant, it will range from golden yellow to black according to its source. Pollen contains many varieties of pigments, of which only a small number have been isolated. Certain pigments are water-soluble, while others are fat-soluble. This accounts for the many varied colors of honey (including the ambers and greens), and the yellow of beeswax is a fat-soluble pigment. Pollen contains the richest known source of vitamins, minerals, proteins, amino acids, hormones, enzymes and fats, as well as significant quantities of natural antibiotics. Most of the known vitamins in pollen exist in perfect proportion, which further enhances their value. Pollen contains 5 to 7 times the amino acids found in equal weights of beef, milk, eggs or cheese. About 35 grams of pollen daily can supply all the body's protein requirements. However, only 25 grams of pollen ingested daily would supply sufficient amounts of each of the essential amino acids. Bee pollen also contains active antibiotic substances that immediately destroy harmful pathogenic bacteria upon contact.

Many universities and colleges throughout the world are discovering the mounting evidence of high performance levels associated with the use of bee pollen. Pythagoras, Hippocrates, Pliny and Virgil all referred to pollen's substantial rejuvenating power, as well as its ability to retard aging. Russia's known centenarians were usually beekeepers, whose diets included large amounts of "scrap" honey, which is a pollen-saturated honey residue salvaged from the bottom of beehives.

Chemical analysis of honey confirms that it consists of two simple sugars - levulose and dextrose. The level of levulose influences honey's tendency to form crystals. The higher the level of levulose makes the honey crystalline. It can be brought back to normal simply by heating. Interesting fact is that honey never freezes! Honey also has the ability to maintain moisture.

Bee pollen has a dramatic effect upon mental perception during athletic performances. The I.Q.'s of children have been doubled during documented clinical tests, and resistance to stress has been significantly increased in both animals and humans.

In addition, bee pollen also contains lecithin, amines, nuclein, guanine, xanthine, hypoxanthine, vernine, waxes, gums, resins, hydrocarbons (0.57%), sterols (0.6%), polypeptides, DNA, ribose, desoxyribose, hexuronic acid, vegetable oils (5% average) and various growth factors.[54]

To make one pound of honey, workers in a hive fly 55,000 miles and tap two million flowers. In a single collecting trip, a worker bee will visit between 50 and 100 flowers and return to the hive carrying over half its weight in pollen and nectar.

Bees are the favorite meal of Merops apiaster, the bee-eater bird. Other common predators are kingbirds, mockingbirds, beewolves, and dragonflies. Solitary bees typically produce neither honey nor beeswax. They are immune from acarine and Varroa mites, but have their own unique parasites, pests and diseases. Solitary bees create nests in hollow reeds or twigs, holes in wood, or, most commonly, in tunnels in the ground. Providing nest boxes for solitary bees is increasingly popular for gardeners. Solitary bees are either stingless or very unlikely to sting

Solitary bees are important pollinators, and pollen is gathered for provisioning the nest with food for their brood. Often it is mixed with nectar to form a paste-like consistency. Some solitary bees have very advanced types of pollen-carrying structures on their bodies. A very few species of solitary bees are being increasingly cultured for commercial pollination.

[54] http://beepollen.info/

Most of these species belong to a distinct set of genera, namely: carpenter bees, sweat bees, mason bees, polyester bees, squash bees, dwarf carpenter bees, leafcutter bees, alkali bees, digger bees.

Currently, a condition known as Colony Collapse Disorder is causing bee populations to decrease dramatically, which also puts our food sources at risk. Researchers attribute this colony collapse to global warming, pesticide use, habitat loss and harmful parasites in the environment. In the United States alone, more than 25 percent of the managed honey bee population has disappeared since 1990.[55]

Without bees to spread seeds, many plants—including food crops—would die off. Bees are responsible for pollinating more than $15 billion each year in U.S. crops. They also produce about $150 million in honey annually. The impact of fewer bees will see the economy take a hit. The global economic impact of the declining bee population will include lower crop yields as well as increased production costs. Some say these numbers will reach upwards of $6 billion a year. So you can see that it is imperative that strong measures be taken to maintain the world's ability to provide high-quality produce and at the same time keep our agricultural sector running well. If something isn't done, there may not be enough honey bees to meet the pollination demands for valuable food crops. But even worse, in a world where up to 100,000 species go extinct each year, the vanishing honey bee could be the herald of a permanently diminished planet.

[55] http://www.nrdc.org/wildlife/animals/files/bees.pdf

The Destruction of Earth

With world population predicted to reach nearly 9.5 billion by 2050, the loss of many of our food crops would cause food shortages and consequent famines worldwide. A 2012 United Nations report warned that we are running out of time to ensure that there will be enough food, water and energy to meet the needs of our rapidly growing population. According to their estimates reported in the Financial Post, by 2030, the world will need at least 50% more food at a time when our environment is developing limits to the supplies. The article goes so far as to say that our current global development model is unsustainable. Our resources are under severe strain. Millions of hectares of forests are lost every year. Our rivers and oceans are being over-exploited. And environmental toxins are being released into our air and water supplies throughout the world. This ecological crisis would be far worse than our current financial crisis. Deforestation, soil degradation, polluted air and water, decline of fish and other wildlife, including the collapse of bee colonies which support the fruits and vegetables we need to survive, foretell the demise of the very planet we live on. British former chief scientific adviser, Sir David King believes we are removing the biodiverse resources at rates faster than the world's last extinction. Figures from the Living Planet index of 2005 indicate that global biodiversity has declined by nearly a third since 1970. During that same period, the ecological footprint of the human population nearly doubled. At that rate, humans would need two planets to provide for their wants in the 2030s, two decades earlier than the previous Living Planet report forecast just two years prior.[56]

"Over consumption and over population underlie every environmental problem we face today"

Jacques-Yves Cousteau

Since life began on Earth, five major mass extinctions have led to large and sudden drops in the biodiversity of species. As background, according to wikipedia, these previous mass extinctions are: Cretaceous–Paleogene; Triassic–Jurassic; Permian–Triassic; Late Devonian; Ordovician–Silurian

The Phanerozoic eon (the last 540 million years) marked a rapid growth in biodiversity in the Cambrian explosion—a period during which nearly every phylum of multicellular organisms first appeared. The next 400 million years was distinguished by periodic, massive losses of biodiversity classified as mass extinction events. The most recent, the Cretaceous–Tertiary extinction event, occurred 65 million years ago, and has attracted more attention than all others because it killed the dinosaurs.

Extreme weather events are becoming more frequent and severe. Wildfires, drought, heat waves, flooding and storm surges destroy agriculture and complete ecosystems. According to the World Resources Institute (WRI), eleven extreme weather events cost the United States more than $1 billion dollars each in 2012. The 2013 World Economic Forum identified "water supply crises" as one of the top five global risks. In 1960, freshwater and

[56] http://www.theguardian.com/environment/2008/oct/29/climatechange-endangeredhabitats

coastal aquatic ecosystems contained fewer than 24 reported dead zones. Today, there are more than 500. The WRI reports that over a billion people live in water-scarce regions today, and are concerned that as many as 3.5 billion could experience water scarcity by 2025. Increasing pollution degrades freshwater and coastal aquatic ecosystems. And climate change is poised to shift precipitation patterns and speed glacial melt, altering water supplies and intensifying floods and drought. The impacts of Hurricane Sandy were amplified by sea level rise, and sea level rise due to climate change has doubled the probability of a Hurricane Sandy-like flood compared to 1950. Every region in the United States has experienced an increase in extreme precipitation events since 1958, while the western United States now sees seven times the number of large-scale wildfires it did 40 years ago. And studies show that several recent droughts in the United States are now more likely (i.e. 2011 Texas drought) and more intense (i.e. the 2013-2014 California drought) due to human-induced climate change.

We are looking at resource collapse of every finite resource extracted or exploited or otherwise processed on an industrial scale. While the media and world leaders focus on fossil fuels, the rest of our resources are in a perpetual state of overuse and decline. What does this mean to us? To put it bluntly, you can get by without oil, but can't go more than three days without water, or a couple of weeks without food.

This just touches on the myriad ways Earth may see its demise. What about an Asteroid impact? The UV rays of an exploding Supernova? A Gamma Ray Burst entering our galaxy? Or could we be sucked into the gravitational pull of a Black Hole? What if the sun suddenly decides to die? Or will our universe simply use up all its energy and burn out? What if Dark Energy began expanding too fast and literally ripped our universe apart? Is there a cannibal galaxy larger than our Milky Way lurking about just waiting to gobble us up like our Milky Way is currently consuming the Sagittarius Galaxy today? Or would we suffer the damages of the electromagnetic radiation rays of a Pulsar?

Scientists at NASA have been watching an asteroid named 1950 DA because it is on a path to collide with Earth on March 16, 2880. Supposedly it will come so close to Earth that it could slam into the Atlantic Ocean with a force of nearly 50,000 megatons of dynamite. They theorize that the probability of impact is only 0.3 per cent, however, this risk is 50 per cent more than an impact from any other asteroid. Over Earth's history, asteroids this size and larger have hit the planet. It is believed such a collision ended the reign of the dinosaur some 65 million years ago. Asteroid 1950 DA was discovered on February 23, 1950 and was only visible for 17 days. It reappeared on December 31, 2000–the 200 year anniversary of the discovery of the first asteroid, Ceres. NASA is tracking 1,400 known hazardous large asteroids that could pass close to Earth.

Gamma rays, thought to be due to the collapse of stars called Hypernovae, are the most powerful events so far discovered in the cosmos. It is theorized that if a gamma ray burst occurred close to Earth, it could bathe the planet in lethal radiation that could deplete our ozone layer and kill much of the life on the surface. Even scarier, gamma ray bursts cannot be predicted, so the first warning we would get, which would be about a flash of brilliant blue light lasting several seconds, would be too late. On April 27, 2013 astronomers observed the supernova in real-time as the light arrived from it. The bursts occur on average about one per day, and nearly all are at the edge of the universe. In our own galaxy, one probably occurs once about every 100,000 to 1 million years. And it is often speculated that this may be the cause of many extinctions in prehistoric times.

Black holes of stellar mass are expected to formed, when very massive stars collapse at the end of their life cycle. After a black hole has formed it can continue to grow by absorbing mass from its surroundings. By absorbing other stars and merging with other black holes, super massive black holes of millions of solar masses may form. There is general consensus that super massive black holes exist in the centers of most galaxies. Science has established that the core of the Milky Way contains a super massive black hole of about 4.3 million solar masses.

Australian astronomers believe finding planets outside our own solar system capable of sustaining life should be a top priority. Dr. Charley Lineweaver of the Australian National University calls the search for life on other planets the most confusing yet most important topic scientists will deal with over the next 10 to 20 years. Believing that not exploring habitable planets for humans would be akin to 15th century Europeans not searching the rest of the world. He further states in his report that a habitable planet may not necessarily have life already existing on it. Arguing that the abiogenis habitable zone is much narrower than for life to survive, he is advocating managing an environment in order to sustain life.

Astrophysicist professor Stephen Hawking also believes we must colonize other planets if we are to survive as a species. Hawking fears global warming, nuclear war or a genetically engineered virus could wipe out the Earth.

A New Heaven, A New Earth

Central to the Christian faith is the premise of a New Earth. In the Book of Revelation, God allows John to see the inheritance of His people through his visions of new heavens and a new earth. Not only does this idea of a new heaven and new earth appear in Revelation 21:1, one can find reference to it in Isaiah 65:17, Isaiah 66:22 and 2Peter 3:13. Each of those Biblical verses state that the old Earth will pass away and that a New Earth will be given to us and our children. Conceptually, Christians will receive new bodies and shall have their feet firmly planted upon the New Earth.

Does the Bible portray the New Earth as something totally different from this current earth or more like a renovation of this current earth? The biblical data seems to support the latter idea. Imagine this scenario from 2 Peter *"But the day of the Lord will come like a thief, and then the heavens will pass away with a loud noise, and the elements will be dissolved with fire, and the earth and everything that is done on it will be disclosed. Since all these things are to be dissolved in this way, what sort of persons ought you to be in leading lives of holiness and godliness, waiting for and hastening the coming of the day of God, because of which the heavens will be set ablaze and dissolved, and the elements will melt with fire?" (2 Peter 3:10-12)*. Is this a warning? Is this a call to us to start preparing for our journey to another world? First Thessalonians 17 says *"Then we which are alive and remain shall be caught up together with them in the clouds, to meet the Lord in the air: and so shall we ever be with the Lord."*

In Revelation 20:11 we see that "*the earth and the heaven fled away; and there was found no place for them.*" As the destruction of the cosmos happens, the new creation replaces the present heaven and earth. It is this destruction that prepares the way for the new heaven and earth as we see in Revelation 20:1, "*And I saw a new heaven and a new earth: for the first heaven and the first earth were passed away; and there was no more sea.*" In the English interpretation we see this as a "second" earth created subsequent to the "first" earth. However, in the Greek interpretation, a new earth (kainos gae) means a different kind of earth than our present earth and heavens. In the Greek, this suggests that the present heaven and earth are temporary and destined to perish. And that the new heaven and earth belong to an age to come and are eternal, never to perish.

The Apostle John describes the new heavenly city as having a great high wall with twelve gates and twelve angels at the gates. And on those twelve gates are written the names of the twelve tribes of the children of Israel. Verse thirteen and fourteen further describe the scene. *"On the east three gates; on the north three gates; on the south three gates; and on the west three gates. And the wall of the city had twelve foundations, and in them the names of the twelve apostles of the Lamb."* The fact that there are gates at each of the four points of the compass suggests the universal nature of the kingdom of God, leading us to believe that the New Jerusalem, or New Earth, will include people from every race and tribe and culture under heaven. According to the Bible, in the new world, the New Eden, all the nations will be joined together as one people. It goes on to say that we will no longer be divided by race, language, culture or social status. Sounds like perfection, doesn't it.

So if we are to believe the Bible, then God is preparing a new world for us. Are we seeing signs of Earth's destruction in our own time? Or will this be years, centuries, or eons away? With our resources being depleted, many nations at war, nations of people living in poverty and more, it does give one something to consider. And scientists are searching the universe for planets which may be able to sustain life.

Using NASA's Kepler Space Telescope, astronomers have discovered the first Earth-size planet orbiting a star in the "habitable zone" -- the range of distance from a star where liquid water might pool on the surface of an orbiting planet. The discovery of Kepler-186f confirms that planets the size of Earth exist in the habitable zone of stars other than our sun. Kepler-186f resides in the Kepler-186 system, about 500 light-years from Earth in the constellation Cygnus. This confirms that planets the size of Earth do exist within the habitable zone of stars other than our sun. While it is yet to be seen whether this is a viable option for habitation, it is a step in the right direction. As we continue to grow in population, using up valuable and necessary resources, it is imperative to discover new worlds and plan for the transition to them in an effort to preserve humanity.

But how will we get there when we do find the New World? These planets are light years away from Earth. Ever since the 1930s physicists have theorized about the existence of wormholes. These theoretical passages through space-time could create shortcuts for long journeys across the universe. Physicists Albert Einstein and Nathan Rosen used the theory of general relativity to propose the existence of "bridges" through space-time in 1935. Although none have been discovered to date, Einstein's theory of general relativity mathematically predicts the existence of wormholes. Scientists continue to explore the concept as a method of space travel hoping that technology will one day be able to use worm holes for space and time travel.

TRUE science and religion are not in conflict. Since antiquity, the topic of the relationship between science and religion have been debated by philosophers, theologians, scientists, and others. Science uses the methodology of empiricism, reasoning and evidence while the methodology of religion uses revelation, faith and sacredness. According to wikipedia, most scientific and technical innovations prior to the Scientific revolution were achieved by

Science, and particularly geometry and astronomy, was linked directly to the divine for most medieval scholars. The compass in this 13th-century manuscript is a symbol of creation.

societies organized by religious traditions. Much of the scientific method was pioneered first by Islamic scholars, and later by Christians. Hinduism has historically embraced reason and empiricism, holding that science brings legitimate, but incomplete knowledge of the world. Confucian thought has held different views of science over time. Most Buddhists today view science as complementary to their beliefs.

Religions have also been the key participants in the development of modern universities and libraries; centers of learning & scholarship - whether pagan, Muslim, or Christian. According a review of Nobel prizes awarded between 1901 and 2000 reveals that (65.4%) of Nobel Prizes Laureates, have identified Christianity in its various forms as their religious preference. Catholic schools have included all manner of scientific study in their curricula for many centuries.

A fundamental principle of the Bahá'í Faith is the harmony of religion and science. Bahá'í scripture maintains that true science and true religion can never be in conflict. The son of the founder of the religion, `Abdu'l-Bahá, stated that religion without science is superstition and that science without religion is materialism. He also urged that true religion must conform to the conclusions of science. Hindu scholastic activities and ancient Indian scientific advancements were so interconnected that many Hindu scriptures are also ancient scientific manuals and vice-versa. Samkhya, the oldest school of Hindu philosophy prescribes a particular method to analyze knowledge. According to Samkhya, all knowledge is possible through three means of valid knowledge:

Pratyaksa or Drstam – direct sense perception,

Anumāna – logical inference and

Śabda or Āptavacana – verbal testimony.

Nyaya, the Hindu school of logic, accepts all these 3 means and in addition accepts one more - Upamāna (comparison).[58]

Nature is not seen as a separate entity in the Muslim religion, but as an integral part of Islam's holistic outlook on God, humanity, and the world. The Islamic view of science and nature is continuous with that of God and religion.

[58]Sarma, Deepak (2011) "Classical Indian Philosophy: A Reader" p.167 Columbia University Press, Samkhya Karika, śloka4

According to Jain doctrine, the universe, soul, matter, space, time, and principles of motion have always existed (a static universe similar to that of Epicureanism and steady state cosmological model). Everything is governed by universal natural laws. It is their belief that it is not possible to create matter out of nothing and therefore the sum total of matter in the universe remains the same (similar to law of conservation of mass). And they further believe that the soul of each living being is unique and uncreated and has existed since the beginning of time.

To quote Stephen J. Gould, if God made each of the half-dozen human species discovered in ancient rocks, why did he create in an unbroken temporal sequence of progressively more modern features—increasing cranial capacity, reduced face and teeth, larger body size? Did he create to mimic evolution and test our faith thereby?

Duane Gish, the leading creationist author, says: "We do not know how the Creator created, what processes He used, for He used processes which are not now operating anywhere in the natural universe.... We cannot discover by scientific investigation anything about the creative processes used by God."

The geological record of fossils follows a single, invariant order throughout the world. The oldest rocks contain only single-celled creatures; invertebrates dominate later strata, followed by the first fishes, then dinosaurs, and finally large mammals. If, according to creationist theory, God created all creatures at once, how did their fossil remains get sorted into an invariable order in the earth's strata?

The story of humanity is written in our genes, and thanks to modern science and technology, we are finally able to read it. (Intelligent Life Magazine 2009). The Genographic Project aims to take 100,000 DNA samples from indigenous people worldwide to establish mankind's path out of Africa. According to the Genographic Creed, every single non-African person is descended from one or two groups of people who traveled across the Red Sea at the narrows of the Bab el-Mandeb, also known as the Gate of Tears, some 50,000 years ago. 99% of the human genome is shuffled from one birth to the next. The Genographic Project intends to track the 1% that is not – the Y-chromosome of the paternal line and the mitochondrial DNA (mtDNA) from the maternal line. Humanity's mtDNA appears to go back to a single woman who lived on the African savannah about 150,000 years ago. While the surviving Y-chromosome leads to a single man living in the Rift Valley of Kenya or Tanzania 59,000 years ago. As society is becoming more homogeneous through generations of multicultural relationships, ancient bloodlines are almost nonexistent, making it more difficult to trace the origins of man.

Genetics may be able to tell us where we come from, but not why we are here nor where the future may lead us. The answer to this great mystery will be the quest of mankind for centuries to come. And it is incumbent on each and every one of us to search our soul and listen to our spirit so that we may discover why we were put here on this Earth.

From the beginning of time, man has searched for the answers to the fundamental questions of life: "Who are we and why are we here? Why do we exist? What is our destiny?

Why were we created? Do we just exist for a short time and then disappear forever? What sets us apart from the animal world? What makes us different from any of the other living creatures?"

The knowledge of one's origin and the reason for one's existence are important things that every person must know, be they rich or poor, learned or illiterate, wise or fool. We believe that this knowledge will free us from thoughts of power, fame, riches or wisdom and that it will lead us to an existence worthy to our higher power.

The answer to life and the universe can't be found by looking through a telescope or examining the tree frogs of the rain forest. It lies much deeper.

"We are here," writes the paleontologist Stephen Jay Gould, "because one odd group of fishes had a peculiar fin anatomy that could transform into legs for terrestrial creatures; because comets struck the earth and wiped out dinosaurs, thereby giving mammals a chance not otherwise available."[59]

Dorion Sagan, the son of Carl Sagan says, "The idea of purpose is usually avoided in science. It seems to imply a mystic reverse causation that cannot be scientific. The reticence is that effects in the material world flow from a prior cause to a future event, while effects in the realm of purpose, which seem to belong only to the realm of the mind, explain present activities in terms of goals not yet achieved. We don't mind attributing such behavior to us, because that's how it feels, but we are stingy to see its existence in others, even the "higher" animals (e.g., similar to us). But the truth is, purposive behavior is everywhere in nature and doesn't require brains. An amoeba swims toward nutrients, a bacterium toward sulfur or oxygen. Sunflowers follow the sun."[60]

[59] http://www.salon.com/2013/06/02/why_are_we_here_evolutions_dirty_secrets/

[60] Cosmic Apprentice: Dispatches From the Edges of Science" by Dorion Sagan. Published by the University of Minnesota Press. Copyright 2013 by Dorion Sagan.

Primordial Soup, Adam

Perhaps the technology already exists that will transport us to a new world. Perhaps the Mayan drawings showing men in what looks like space ships are real. No one knows, but we will continue to question and search for answers.

We have looked at centuries of scientific and religious study regarding the origins of life. And it seems that even with all the research we are no closer to finding the real answer to how life began. Could it be that we did arise from an amoeba in the primordial soup? Perhaps we were a speck of bacteria that somehow became trapped in a crystal geode. Could it be that within that same geode existed a honeycomb filled with bee larva that could have provided the necessary components to synthesize into a life form? Not a full blown human but something less complex that eventually evolved into something else. Could it be that the electromagnetism from the crystals have provided the energy catalyst to spur the development of life? Who is to say? What do we really know? Who put the first plant or the first bee on Earth? For that matter, where do rocks and crystals come from? Silica of course! But where does the silica come from? And how does silica have the intelligence to replicate itself? Scottish chemist A. G. Cairns- Smith proposed that crystals were the first living organisms. Crystals have great copying fidelity. Heredity in crystals appears to be a simple, natural and common occurrence. This makes crystals appear to be the most likely self-replicating agents in a pre-biotic environment. And crystals grow best in a very saturated environment. While Cairns-Smith's theory has been greatly neglected to date, it still seems that his clay (crystal) hypothesis has merit.

C. Jeffrey Brinker, Fellow, Sandia National Laboratories, Distinguished and Regent's Professor, the University of New Mexico, Albuquerque, NM USA has discovered a process he calls Silica Cell Replication in which mammalian cells direct their exact replication in silica. There is nothing closer to carbon than silicon in terms of its chemistry. Like carbon, silicon forms four bonds. CH4 is methane and SiH4 is silane. Since carbon and silica are analogous molecules could it be possible for life to have formed from silicate crystals?

Silicon is an essential mineral for the human body and has a primary function in helping develop and maintain structural and functional integrity of the connective tissues as well as the skeletal system. It is an essential trace mineral that our bodies need for stronger bones and joints as well as supple skin. The actual functions of silicon in the body are not thoroughly understood. But it is curious that it plays such an important part in the very architecture of the human body as well as being one of the most abundant minerals on earth. It's a fact that it can be found in foods like apples, onions, peanuts, raw cabbage, pumpkin, cucumber, HONEY, almonds, oats, oranges and fish. Is there a connection? I think so.

My theory is that the Big Bang rained down debris from the expansion depositing minerals and amino acids, the building blocks of protein, upon a barren, hostile environment and that the heat and evaporation from the cooling formed water to create an environment suitable for simple life to form. My theory has all the connections—Multiverse, Big Bang, Amino Acids, Complex Life and a Higher Power that is God.

Compartment in a crystal filled geode within which lie all the necessary components to create life. Includes the intelligent DNA of bees, the Deoxyribose sugar necessary for life that are found in honey, and the mathematical genius of the honeycomb. And the universal creation code.

Creation Code

Rolando Rivero

$$O = O^{12}\ \Delta^{20}$$

12 + 20 + 1 = 33

Crystals 32 + 1 = 33

Bee chromosomes 32 +1=33

Sugar DNA 8+8+8+8=32=1 double helix.

Rolando's Adam formed inside a compartment within a crystal geode containing all the necessary building blocks for life to begin. The Garden of Adam and Eve. The Garden of Eden!

Genesis 2:8

The LORD God planted a garden toward the east, in Eden; and there He placed the man whom He had formed.

The God Particle

From rather small molecular building blocks nature uses the process of self-assembly to build large functional structures — like cell membranes, protein aggregates and the DNA double helix. This self-assembly process may provide a method to create complex nanostructures that could open the doors to new technologies leading to the development of next-generation materials with favorable electronic properties like carbon nanotubes or nanowires. According to a report in Nature Nanotechnology, researchers have been able to get engineered "DNA origami" to self-organize on silicon. They say that several groups of researchers have demonstrated that DNA itself can be used to store or manipulate data, and also that by juggling DNA in a test tube or within bacteria it is able to solve simple calculation tasks. Current methods leverage the ability to design DNA strands into regular shapes such as triangles. These structures are tiny and quite difficult to work with, but the chemical groups hanging off of DNA molecules could be used as anchor points for them.

Built by the European Organization for Nuclear Research (CERN) between 1998 and 2008, the Large Hadron Collider is the world's largest and most powerful particle collider. More than 10,000 scientists and engineers from more than 100 countries, in addition to hundreds of universities and laboratories all worked together to develop the Hadron Collider which is underground in a tunnel between the French and Swiss borders. It was shut down in February of 2013 to upgrade it so that when it comes back on line in late 2014, it will operate at double its current maximum energy.

In 2012 researchers at the CERN particle accelerator in Geneva, Switzerland announced that they had discovered a subatomic particle– the Higgs boson, or God particle. The particle is named for Dr. Peter Higgs, one of six physicists who first described it in 1964. This particle may explain why some fundamental particles have mass when the arrangement controlling their interactions should require them to be massless, and why the weak force has a much shorter expanse than the electromagnetic force.

Outside the scientific community, the Higgs boson is regularly referred to as the "God particle". This nickname came from the title of a 1993 book on the Higgs boson and particle physics written by Nobel Physics prizewinner and Fermilab director Leon Lederman.

"Today ... we have the standard model, which reduces all of reality to a dozen or so particles and four forces. ... It's a hard-won simplicity [...and...] remarkably accurate. But it is also incomplete and, in fact, internally inconsistent... This boson is so central to the state of physics today, so crucial to our final understanding of the structure of matter, yet so elusive, that I have given it a nickname: the God Particle. Why God Particle? Two reasons. One, the publisher wouldn't let us call it the Goddamn Particle, though that might be a more appropriate title, given its villainous nature and the expense it is causing. And two, there is a connection, of sorts, to another book, a much older one..."— Leon Lederman[61]
Physicists hope that the collider can answer many questions, for example, what is the dark matter that seems to be 27% of the mass-energy of the universe? Also whether there are other dimensions. Perhaps the answers will come with the newly renovated collider.

[61] Leon M. Lederman and Dick Teresi (1993). The God Particle: If the Universe is the Answer, What is the Question. Houghton Mifflin Company.

Righteousness and Procreation

God the Creator of the multiverse, our universe, the Big Bang, our Milky Way, our solar system, Earth, the creator of everything. God created everything because of his best creation, human beings. Human beings are at the center and the purpose of everything, 214 billion years and 1.371196 quintrillion multiverse space size in the making.

There are Angels throughout our multiverse, even on earth before human beings were created. These angels were almost perfect. Why did God create human beings, his perfect creation before our tangible thinking? When I say before our thinking it is because it's obvious most people in this world are righteous, a small percentage or a few are evil.

You see God created us so perfect for us to do our own righteous things as well. But why did God take a chance in creating us when there were almost perfect angels? There is a reason why I say almost perfect angel; it is because God is the father of everything. God is the Father of the angels. God is the Father of humans. Angels are God's creations and humans are his children. These angels and children love God very much but there is something missing. I am a father and I love my children very much. I would give my life for my children and my children love me. My father loves me more than I love him. I love my children more than they love me. You see, you have to be a father to understand that extra love. My children will not understand this kind of love until they have their own children. Even though humans have this love for their children, humans also have love for God. God is our father and we are his children. Angels cannot have children. That is the missing love link. God loves his angels, but his Angels cannot bear children, so they cannot be fathers. Angels just don't understand what it means to bear children and nor do they understand the depth of a father's love for his children. That, in my opinion, is why God took a chance in creating humans all in the name of full and complete love. Pro-creation.

So between mapping the multiverse and years of perfecting God's pulsating energy creation of a multiverse, the love transformation between angels and human begins one last phase and is the reason for perfecting the multiverse and God's creation.

I know everyone in this world has God given talents. Some people are in the field of religion, some are in the scientific field, some are in the excellence of living. Some people are specially chosen from God. And people learned from these special children from God. We learned from our ancestors. Most of our ancestors in spirit form are with our God creator. It is a continuous circle from our past, present and future. This generation will show the next generation and the next generation will learn from us. I know that the next generations to come will make changes. This world was created for humans but unfortunately the world resources will start to decline requiring humans to find other new resources with new inventions and new ideas. These new inventions and new ideas will require outer space travel. It's time that humans start producing new technology for outer space travel.

That's why I know my vision for multiverse mapping is a crucial step for mankind. We need new technology to create vessels that can travel far into outer space. That's why we need the world minds to create such an empire of new thoughts because space travel requires generations. By perfecting and dissecting every angle of my multiverse map there will be discoveries of neutral areas for speedy travel. These neutral areas are the pushing and pulling inbetween the multiverse and in the center of the points are the speedy travel areas. These areas would be easier to find once we are out there. These areas are triangular. Triangles are a universal mathematical language for people in speedy travel when people have perfected long-term human vessels travel and are ready to travel far into outer space.

Perhaps we have received a message regarding how we will travel to a new earth from 1 Thessalonians 4:13-18: *"But we do not want you to be uninformed, brethren, about those who are asleep, so that you will not grieve as do the rest who have no hope. 14 For if we believe that Jesus died and rose again, even so God will bring with Him those who have fallen asleep in Jesus. 15 For this we say to you by the word of the Lord, that we who are alive and remain until the coming of the Lord, will not precede those who have fallen asleep. 16 For the Lord Himself will descend from heaven with a shout, with the voice of the archangel and with the trumpet of God, and the dead in Christ will rise first. 17 Then we who are alive and remain will be caught up together with them."* The words "caught up" come from the Greek word *harpazo*. It means to seize, to pluck, to pull, to catch away, to catch up, to take by force, to take for one's self eagerly, to rescue from the danger of destruction. There are six people in the Bible who have been "caught up." They are Enoch, who did not die but was taken up. (Genesis 5:24 and Hebrews 11:5). Elijah - 2 Kings 2:11 who when walking along with Elisha went up by a whirlwind to heaven leaving Elisha on earth. Then the prophet Ezekiel who 300 years after Elijah's rapture was lifted up by the spirit into the inner court. (Ezekiel 43:5) Phillip, the deacon, was also supernaturally transported by God. (Acts 8:39-40) The

Apostle Paul recorded his experience in 2 Corinthians 12. And of course, Jesus was lifted up after his resurrection. ENOCH was TAKEN UP. ELIJAH was CAUGHT UP. EZEKIEL was CARRIED BY THE SPIRIT. PAUL was CAUGHT UP to the third heaven. PHILLIP was transported. JESUS was "caught up" or "raptured" into heaven at his ascension.

The last phase, God's creation 214 billion years and 1.371196 quintillion space size in the making of our multiverse, universe, solar system, Earth. My story has just begun for this book barely touches on my theories that will change how we think about the creation of our universe. My hope is that you will begin to understand that neither science alone nor religion alone are the sole authority of how we got here. Nor does either truly hold the answers for where we will go from here. It is time for both communities to work together for the good of mankind.

When we can open our minds to new ways of thinking and when science and religion can see that they each have an important role to play in providing the answers we seek, there will be a great release of knowledge that will satisfy the quest for how the universe really began.

Every righteous person has a role to play with me creating the secrets of my multiverse whether in the past, present or future, whether knowing or not knowing, whether directly or indirectly. For those who are not righteous persons, who are bad and evil you have no role whatsoever in being part as Psalms 104:35 says, "Let the sinners be consumed out of the earth, and let the wicked be no more." Evil people or Antichrist people have no part in outer space universe travel. If they think of infiltrating an outer space vessel or universal travel they are completely wrong. Remember they will be on their own because the Antichrist is Earth bound – jailed. They will have no backing of the Antichrist whatsoever. God is the creator of all universes and will strike them down in a heartbeat as the Bible promises. There is no room for bad or unrighteous people. There is going to be a time when the world is at peace. God knows when people will be ready. Then God is going to send his right-hand angel to our Heaven to guide us into outer heavens. But, using the world's new and improved multiverse map and the guide of God's right-hand angel, we humans will be ready and in motion to travel the heavens of heavens our universe and the multiverse.

The Bible says that "***No man hath seen God.***" Does this mean at anytime? Or will we see him in the future? Why is this even in the Bible? What does it mean? Humans are the perfection of everything God has created. The day will come when God our father will meet humans, His children in the living flesh. And He will embrace His perfect creation with a big strong hug and say finally, "I love you my children." In the meantime, I, Rolando Rivero, a child of God, will walk with Him, my father God, in my heart every minute of the day. Thank you God, Jesus Christ, and the Holy Spirit with all my heart and soul for the opportunity to share my story with like-minded people who are searching for answers to the many questions surrounding the Creation.

Heaven One, Heaven Two and Heaven Three

The description of heaven cannot even enter in the minds of people.

Traditional Thinking on Heaven

Heaven 1, Earth's Atmosphere in the middle where the birds and clouds are.

Heaven 2, Outer Space, the sun, stars and moon

Heaven 3, God's Dwelling Place

Rolando's Theory

My theory is that Heaven is all of the universe. That when we receive a new Heaven and new Earth, it will be outside of our current universe, first Heaven (blue). The second Heaven (yellow zone - 12 multi-verses) is within the multiverse, but outside our vast universe. This is where the 12 apostles and the 12 gates will be. God lives on the outer edge of the third Heaven (green zone - 20 multi-verses) just outside our reach and understanding. "And God made the firmament, and divided the waters which were under the firmament from the waters which were above the firmament: and it was so." (Genesis 1:7)

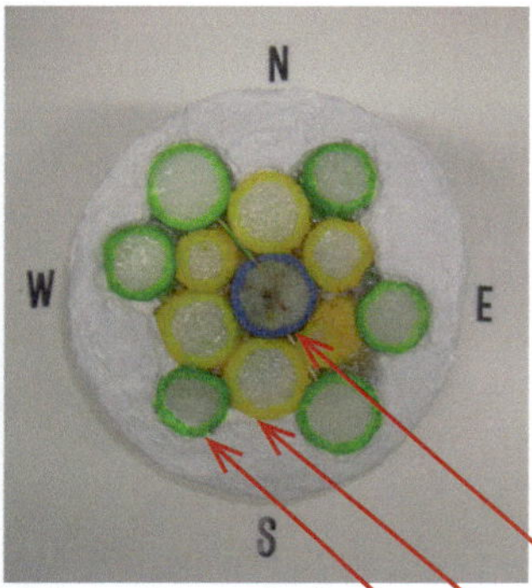

White area is God's light of the lamb, God's energy encompassing all of the multiverses mapped.

Heaven 1, The exit side of our universe (blue)

Heaven 2, The exit side of yellow zone 2 -12 multi-verse

Heaven 3, The exit side of green zone 3 - 20 multi-verses

In the bible Paul says he was "caught up" to the third Heaven. He was referring to the dwelling place of God. The Mormons teach that the three heavens consist of telestial, terrestrial, and celestial. They divide them into compartments dwelt by people after they die. The fact that Jesus passed through the heavens gives evidence there is more than one heaven. "*Seeing then that we have a great High Priest who has passed through the heavens, Jesus the Son of God, let us hold fast our confession*" *Hebrews 4:14*. The Scripture also speaks of heavenly spheres beyond what is visible from earth. It is called the heaven of heavens. "*Indeed heaven and the highest heavens belong to the Lord your God, also the earth with all that is in it*" *Deuteronomy 10:14*. "*Praise Him, you heaven of heavens and you waters above the heavens!*" *Psalm 148:4*.

The ultimate purpose of His creation is to choose a group of obedient people to live with him, to enjoy his glory forever. "*Isaiah 64:4 For since the beginning of the world men have not heard, nor perceived by the ear, neither hath the eye seen, O God, beside thee, what he hath prepared for him that waiteth for him. 5 Thou meetest him that rejoiceth and worketh righteousness, those that remember thee in thy ways:"*

There is a city in the new earth: the New Jerusalem. Revelation says that in the New Jerusalem, God "will dwell with them, and they will be his people, and God himself will be with them and be their God." Rev 21:4 As a result, there is "no temple in it, for the Lord God Almighty and the Lamb are its temple." Nor is there a need for the sun to give its light, "for the glory of God illuminated it, and the Lamb is its light" Rev 21:22-23.

The first day of the biblical creation narrative begins with darkness, into which is introduced the creation of light, and the separation of this light from the darkness (as distinct from the creation of the sun and moon on the fourth day of creation). God's true light then opened in the center and gave space to darkness first. So our kind of light can travel. Darkness is the key. Light uses darkness to travel and darkness is faster everywhere. The city will also be a place of great peace and joy, for "*God will wipe away every tear from their eyes. There will be no more death, nor sorrow, nor crying; and there will be no more pain, for the former things have passed away" Rev 21:1-4.*

The city itself has a large wall with twelve gates in it that are never shut, which have the names of the twelve tribes of Israel written on them. Each of the gates is made of a single pearl, and there is an angel standing in each one. The wall also has twelve foundations adorned with precious stones, and on the foundations are written the names of the twelve apostles. The gates (tribes of Israel) and foundations (apostles) are often interpreted as symbolizing Israel and the Church. Could these 12 gates be my 12 yellow multi-verses?

The city has a river that proceeds "*out of the throne of God and of the Lamb." Rev 22:1.* Next to the river is the tree of life, which bears twelve fruits and yields its fruit every month. The last time the tree of life was mentioned in the Bible was in connection with the Garden of Eden.[Gen 2:9] God drove Adam and Eve away from it because it bestowed eternal life and he did not want them to have it in their degraded state[Gen 3:22] In the New Jerusalem, the tree of life reappears, and everyone in the city has access to it. Genesis also tells us that the earth was cursed because of Adam's sin,[Gen 3:17] but John writes that in the New Jerusalem, *"there will be no more curse." Rev 22:3.*

God's Signs

Regardless of where we came from, where we may go from here you can be assured that there is a plan for tomorrow. Our very human nature is geared for survival and for those who place their faith in a higher power the Christian Bible states that "*For I know the plans I have for you, declares the Lord, plans for welfare and not for evil, to give you a future and a hope.*" *Jeremiah 29:11.*

So many mysteries, so many theories, so many belief systems...yet we are one people inhabiting this one planet filled with many miraculous things...all here for our pleasure and discovery and study. May it be that you,too, just as I have done, question and research and share your discoveries and theories about who we are, how we got here, why we are here, where we are going and how we are changing as a species, as a world. May you also keep your eyes open for signs and evidence that may provide the answers mankind has been searching for.

Al-Hazen discovered how images are formed by light, but that was far from his greatest achievement. Al-Hazen was the first person ever to set down the rules of science. He created an error correcting mechanism. A systematic and relentless way to sift out misconceptions in our thinking. Each and every day new discoveries are being made. Every new generation challenges the thoughts and theories of those previous to them. Intelligent, inquiring minds will not rest with the status quo. The thoughts and theories in my book are presented to you as catalysts for creative thinking. Here to raise your awareness of possibility. Here to cause you to question the current thoughts and theories of accepted scientific proof of how our universe began. Remember that the world knew everything was made of atoms long before the technology existed to provide actual scientific proof that this was the truth. How many other things has our ever expanding technological knowledge proved and disproved? The only thing that is certain is that we must continually renew our minds and search onward for the truth. Finding truth is difficult and the road to it is rough, as seekers after truth you will be wise to withhold judgment, and not simply put your trust in the writings of the ancients. You must question and critically examine those writings from every side. You must submit only true arguments and experiments and not to the saying of any person, for every human being is vulnerable to all kinds of imperfection. As seekers after truth we must also suspect and question our own ideas as we perform our own investigations to avoid falling into prejudice or careless thinking.

Do not be conformed to this world, but be transformed by the renewal of your mind, that by testing you may discern what is the will of God, what is good and acceptable and perfect. Romans 12:2.

Take this course and the truth will be revealed to you. *And you will know the truth, and the truth will set you free. John 8:32.*

May you share this quest with your children and them with theirs, for we know not how long the Earth will last. And perhaps, through this everlasting search for the Exit Side of the Universe, the future will be prepared for the journey to a New Earth and worlds beyond our own time and understanding.

You may have noticed that there is a four-letter word at the top of each page of my book. It is a very important word and one you may not have seen or heard of before but you already know it.

YHVH is life. YHVH is light (energy/matter). YHVH is truth. You already know God's name. Oh, yes, there are many, many names people have given to God, but He picked one and only one for Himself. God's Real name is the same for all people everywhere. It matters not at all which religion or cultural heritage you are from! Truth transcends all boundaries we seek to erect. And no matter who you are or where you are from, you have known this Name from the day your mind first began to awake. You see it and hear it each and every day.

The personal name of God, YHVH, I am Who I am, embodies all matter, energy, life, and truth. YHVH, the I Am that I Am, the creator of the universe, depends only on Himself for existence. He is the source and origin of all that exists. In Him we live and move and have our being.

And Moses said unto God, Behold, when I come unto the children of Israel, and shall say unto them, The God of your fathers hath sent me unto you; and they shall say to me, What is his name? What shall I say unto them?

And God said unto Moses, I AM THAT I AM: and he said, Thus shalt thou say unto the children of Israel, I AM hath sent me unto you.

And God said moreover unto Moses, Thus shalt thou say unto the children of Israel, The LORD God of your fathers, the God of Abraham, the God of Isaac, and the God of Jacob, hath sent me unto you: this is my name for ever, and this is my memorial unto all generations. Exodus 3:13-15

To translate means to turn from one language into another or from a foreign language into one's own, while transliterate means to change (letters, words, etc.) into corresponding characters of another alphabet or language. To transliterate is to spell a word using the letters of another language.

The English transliteration of God's personal name is YHVH.

The four Hebrew letters transliterated YHVH are:

י a Yod, rhymes with "rode", which we transliterate "Y"

ה a He, rhymes with "say", which we transliterate "H"

ו a Vav, like "lava", which we transliterate "W" or "V"

ה another He

In ancient times the Hebrew language did not use vowels. And was written from right to left. Translated into English and written from left to right with vowels added, YHVH becomes YAHWEH. "I am" is the English translation of the meaning of God's personal name.

YHVH

Imagine yourself in a pitch dark room. You turn on the switch and light floods the room. An actual energy appears. Photons (what light is made of) stream out of the bulb and illumine the room. Switch to off and all the photons disappear. Nothing different is now coming out of the bulb "darkening" the room the way the photons were lighting it. The photons are just gone!

No one on the earth knows what light is. We know it moves in waves and we know that it is consists of particles, and we know that particles can't move in waves and that waves can't contain particles. But that's what light is. And whatever light is, it is!

"I AM THAT I AM." God is light, and in Him there is no darkness at all! YHVH is life. Life, whatever it is, is. Death is simply the absence of life.

YHVH is light (energy/matter). Energy is. Light is only one form of energy. Matter is. We know that energy and matter are opposite sides of the same coin. We can change matter to energy and vice versa; however, the total amount of energy/matter can not be changed. It is. And it is YHVH.

YHVH is truth. Truth states what is. Falsehood states what is not.

Sometimes YHVH is translated as the self-existent one. YHWH, the I Am that I Am, the creator of the universe, depends only on Himself for existence. He is the source and origin of all that exists. In Him we live and move and have our very being.

All of YHVH's nature and attributes are embodied in His name. He is the ultimate ground. We are the figures. He is infinite subject. We are object. He is the source. We the proceeds. God, the creator of the universe, is. And that is-ness is not dependent on any other reality.

YHVH is everywhere. YHVH has revealed His name to us. His life is now an open book, ready for whoever wants to read it. YHVH has revealed His name to us. What is more, He has given it to us to use. YHVH's name is His image. It presents Him to the world. You are YHVH's name. You are created in the image of YHVH. This is not chance. This is not a game. This is not something that has been thought of recently. It is the oldest of revelations And the most powerful. As YHVH is reflected in the human body, The Yod is the head, the 1st He is the arms, The Vav is the torso, The 2nd He is the legs.

YHVH is reflected in us. The Yod is our mind, The 1st He is our emotions, The Vav is our life energy, The 2nd He is our body. YHVH is reflected in life. The Yod is our intellect, The 1st He is our desire, The Vav is our animation, The 2nd He is our matter. YHVH is reflected in our world, The Yod is mankind, The 1st He is the animal kingdom, The Vav is plant life, The 2nd He is our physical universe YHVH is reflected in the cosmos. The Yod is the creator/consciousness of creation. The 1st He is the impulse of creation, The Vav is the immaterial created universe, The 2nd He is the material created universe. This one name reflects and echoes itself in all dimensions.

May you recognize His Name radiating within you. May you recognize His Name radiating around you. May your heart, mind and eyes be opened to the greater understanding that is available to each of us through the constant renewal of our minds and the search for truth. May it be revealed to you.

"And do not follow the customs of the present age, but be transformed by the entire renewal of your minds, so that you may learn by experience what God's will is--that will which is good and beautiful and perfect." (Romans 12:2)

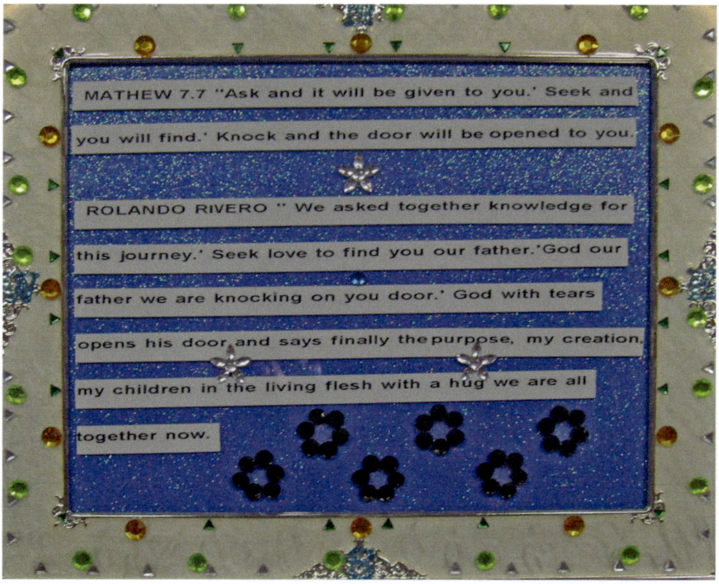

1 *Bless the LORD, my soul! LORD, my God, you are great indeed!*
You are clothed with majesty and splendor,

2 *robed in light as with a cloak. You spread out the heavens like a tent; a*

3 *setting the beams of your chambers upon the waters. You make the*
clouds your chariot; traveling on the wings of the wind.

4 *You make the winds your messengers; flaming fire, your ministers.*

5 *You fixed the earth on its foundation, so it can never be shaken.*

6 *The deeps covered it like a garment; above the mountains stood the waters.*

7 *At your rebuke they took flight; at the sound of your thunder they fled.*

8 *They rushed up the mountains, down the valleys to the place you had fixed for them.*

9 *You set a limit they cannot pass; never again will they cover the earth.*

10 *You made springs flow in valleys that wind among the mountains.*

11 *They give drink to every beast of the field;e here wild asses quench their thirst.*

12 *Beside them the birds of heaven nest; among the branches they sing.*

13 *You water the mountains from your chambers, from*
the fruit of your labor the earth abounds.

14 You make the grass grow for the cattle and plants for
people's work to bring forth food from the earth,

15 wine to gladden their hearts, oil to make their
faces shine, and bread to sustain the human heart.

16 The trees of the LORD drink their fill, the cedars of Lebanon, which you planted.

17 There the birds build their nests; the stork in the junipers, its home.

18 The high mountains are for wild goats; the rocky cliffs, a refuge for badgers.

19 You made the moon to mark the seasons, the sun that knows the hour of its setting.

20 You bring darkness and night falls, then all the animals of the forest wander about.

21 Young lions roar for prey; they seek their food from God.

22 When the sun rises, they steal away and settle down in their dens.

23 People go out to their work, to their labor till evening falls.

24 How varied are your works, LORD! In wisdom you have
made them all; the earth is full of your creatures.

25 There is the sea, great and wide! It teems with countless
beings, living things both large and small.

26 There ships ply their course and Leviathan, whom you formed to play with.

27 All of these look to you to give them food in due time.

28 When you give it to them, they gather; when you open your hand, they are well filled.

29 When you hide your face, they panic. Take away their
breath, they perish and return to the dust.

30 Send forth your spirit, they are created and you renew the face of the earth.

31 May the glory of the LORD endure forever; may the LORD be glad in his works!

32 Who looks at the earth and it trembles, touches the mountains and they smoke!

33 I will sing to the LORD all my life; I will sing praise to my God while I live.

34 May my meditation be pleasing to him; I will rejoice in the LORD.

35 May sinners vanish from the earth, and the wicked be no
more. Bless the LORD, my soul! Hallelujah!

Psalm 104

YHVH

YHVH

Further Meanderings

The Holy Bible in any version

The Origin Of The Universe: Science Masters Series by John D Barrow ISBN-10: 0465053149

Science & Religion: From Conflict to Conversation (Crossway Classic Commentaries) by John Haught ISBN-10: 0809136066

Finding Darwin's God: A Scientist's Search for Common Ground Between God and Evolution by Kenneth R. Miller ISBN-10: 0061233501

The Story of Life by Richard Southwood ISBN-10: 0198607865

Wonderful Life: The Burgess Shale and the Nature of History by Stephen Jay Gould ISBN-10: 039330700X

Ever Since Darwin: Reflections in Natural History by Stephen Jay Gould ISBN-10: 0393308189

The Panda's Thumb: More Reflections in Natural History by Stephen Jay Gould ISBN-10: 0393308197

The Mismeasure of Man (Revised & Expanded) by Stephen Jay Gould ISBN-10: 0393314251

A Brief History of Time by Stephen Hawking (Sept. 1, 1998) ISBN-10: 0553380168

The Grand Design by Stephen Hawking & Leonard Mlodinow ISBN-10: 055338466X

The Theory Of Everything by Stephen W Hawking ISBN-10: 8179925919

The Universe in a Nutshell by Stephen Hawking ISBN-10: 0965326942

Black Holes and Baby Universes and Other Essays by Stephen W. Hawking ISBN-10: 0553374117

Biocentrism: How Life and Consciousness are the Keys to Understanding the True Nature of the Universe by Bob Berman & Robert Lanza ISBN-10: 1935251740

The Unobservable Universe: A Paradox-Free Framework for Understanding the Universe by Scott M. Tyson (Author), Catherine J. Rourke (Editor), Thomas M. Hill (Editor) ISBN-10: 0983243808

God Theory, The: Universes, Zero-Point Fields, and What's Behind It All by Bernard Haisch ISBN-10: 9781578634361

Einstein's Cosmos: How Albert Einstein's Vision Transformed Our Understanding of Space and Time (Great Discoveries) by Michio Kaku ISBN-10: 0393327000

Physics of the Impossible: A Scientific Exploration into the World of Phasers, Force Fields, Teleportation, and Time Travel by Michio Kaku ASIN: B002A7MJS2

Physics of the Future: How Science Will Shape Human Destiny and Our Daily Lives by the Year 2100 by Michio Kaku ISBN-10: 9780307473332

The Future of the Mind: The Scientific Quest to Understand, Enhance, and Empower the Mind by Michio Kaku

Parallel Worlds: A Journey Through Creation, Higher Dimensions, and the Future of the Cosmos by Michio Kaku

Pale Blue Dot: A Vision of the Human Future in Space by Carl Sagan

Cosmos by Carl Sagan (Author), Ann Druyan (Introduction), Neil deGrasse Tyson (Foreword) ISBN-10: 0345539435

The Dragons of Eden: Speculations on the Evolution of Human Intelligence by Carl Sagan ISBN-10: 0394410459

Shadows of Forgotten Ancestors by Carl Sagan & Ann Druyan ISBN-10: 0345384725

Science and the Akashic Field: An Integral Theory of Everything by Ervin Laszlo ISBN-10: 1594770425

The Ancestor's Tale: A Pilgrimage to the Dawn of Evolution by Richard Dawkins ISBN-10: 061861916X

The Extended Phenotype: The Long Reach of the Gene by Richard Dawkins ISBN-10: 0192880519

The Blind Watchmaker: Why the Evidence of Evolution Reveals a Universe without Design by Richard Dawkins ISBN-10: 0393315703

The Sixth Extinction: An Unnatural History by Elizabeth Kolbert ISBN-10: 1250062187

The Beekeeper's Lament: How One Man and Half a Billion Honey Bees Help Feed America by Hannah Nordhaus ISBN-10: 006187325X

The Sacred Bee in Ancient Times and Folklore (Dover Books on Anthropology and Folklore) by Hilda M. Ransome ISBN-10: 048643494X

The Shamanic Way of the Bee: Ancient Wisdom and Healing Practices of the Bee Masters by Simon Buxton ISBN-10: 1594771197

Love is in the Earth: A Kaleidoscope of Crystals - The Reference Book Describing the Metaphysical Properties of the Mineral Kingdom by Melody (Author), Julianne Guilbault (Illustrator) ISBN-10: 0962819034

The Crystal Bible by Judy Hall ISBN-10: 1582972400

YHVH

Made in the USA
Las Vegas, NV
04 November 2026

57424661R10079